PAUMANOK PUBLICATIONS

I0394531

PASSIVE COMPONENT
INDUSTRY

Electronic Industries Alliance

An affiliate publication of
Electronic Components Association

The Only Magazine Dedicated Exclusively To The Worldwide Passive Electronic Components Industry

   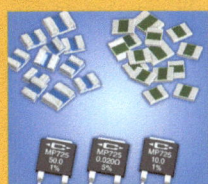 

2009 BUYERS GUIDE
&
RESOURCE DIRECTORY

A Comprehensive Guide to the Passive Component Industry

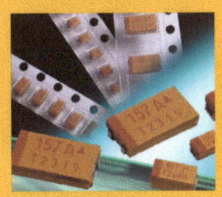    

monday

MONDAY, APRIL 29

Noon - 1:00 pm	Registration
1:00 - 5:00 pm	Summit Program User/Application Developer Presentations

Please see online program at http://ecaus.org/conference/spring_summit_2009/program.htm for program updates.

International Standards

Update	
6:30 pm - 7:30 pm	Welcome Reception (open to all attendees)

• • •

Agendas for committee meetings are posted at **www.ecaus.org/engineering/committees/** *under each committee listing.*

SPRING ENGINEERING SUMMIT

tuesday

TUESDAY, APRIL 28

7:30 am - 4:30 pm	Registration
8:00 - 10:00 am	S-1 Passive Component Committees Steering Group
9:00 am - 5:00 pm	CE-2.0 National Connector Committee
10:00 - 10:30 am	Break
10:00 - 11:00 am	P-14 Overcurrent Protection Devices Committee
11:00 am - Noon	P-10 Integrated Passive Devices Committee
Noon - 1:00 pm	Lunch (open to all attendees)
1:00 - 2:00 pm	P-1 Resistive Devices Committee
2:30 - 3:00 pm	Break
2:00 - 6:00 pm	Tantalum Working Group
7:00 - 9:00 pm	Dinner with Keynote

wednesday

WEDNESDAY, APRIL 29

7:30 am - 4:30 pm	Registration
8:00 - 9:00 am	P-3 Inductive Components Committee
8:00 am - 5:00 pm	CE-2.0 National Connector Committee
9:00 - 11:00 am	STC Soldering Technology Committee
10:00 - 10:30 am	Break
11:00 am - Noon	P-2.5 Tantalum Capacitors Committee
Noon - 1:30 pm	Awards Luncheon (open to all attendees)
1:30 - 3:30 pm	P-2.2 Paper, Film, Mica & Wet-Electrolytic Capacitors Committee
2:30 - 3:00 pm	Break
3:30 - 5:30 pm	Ceramic Working Group

APRIL 27-30, 2009 • NEW ORLEANS, LA

thursday

THURSDAY, APRIL 30

7:30 - 10:30 am	Registration
8:00 am - 4:30 pm	Automated Component Handling Committee
8:00 - 10:00 am	P-4 Mechanical Outlines Committee
10:00 - 10:30 am	Break
10:30 am - Noon	P-2.1 Ceramic Dielectric Capacitors Committee
Noon - 1:00 pm	STPC (Working Lunch)
1:00 - 3:00 pm	P-2.7 Electrochemical Double-layer Capacitors Committee

For more information on attending the 2009 Spring Engineering Summit, please visit our website at www.ecaus.org.

2500 Wilson Blvd., Suite 310
Arlington, VA 22201
703.907.8024

 muRata
Innovator in Electronics

Wonder what lies inside your trusted PDA or handheld device?

Well, Murata ceramic passive products most likely, especially our range of ultra small, high Q, capacitor array and high capacitance MLCC ceramic capacitors.

small Products...

BIG Solutions!

High Q

GJM03/15 series - EIA Size (0201, 0402)
- High Q capacitor series for high frequency applications in the 500MHz to 10GHz range.
- Small size, with voltages of 6.3, 25 and 50Vdc.
- C0G (-55°C to 125°C with 0 ±30ppm/°C).
- Capacitance range of 0.2 to 33pF.
- Great for low power applications: PA modules, W-LAN, wireless modems, remote keyless entry, wireless PDAs, GPS, antenna tuning, and Bluetooth® modules.

Capacitor Array

GNM series - EIA Size (0302, 0504)
- Small 2 element multicap packages with high capacitance up to 2.2uF@10Vdc.
- 0302-size 2-element array has achieved 1.0µF static capacitance.
- Achieves higher static capacitance compared with single element capacitors.
- Reduced mounting surface area and cost due to high-density mounting.
- Applications: Used as a charge pump for LCD phones, mobile PCs, and digital AV equipment.

Ultra Small

GRM02 series - EIA Size (01005)
- Ultra small size 0.4mm x 0.2mm.
- Comparing with 0201 size product, 01005-size product has an area reduction of 45% and a volume reduction of 30%.
- Tailored for miniature applications: PA module, high frequency circuit, Bluetooth® modules, and portable electronics.
- 01005 C0G 0.2pF to 47pF@16Vdc; 56 to 100pF@10Vdc.
- 01005 X5R 680pF to 0.01uF@6.3Vdc.
- 01005 X7R 100pF to 560pF@10Vdc.

GRM03/15 series - EIA Size (0201, 0402)
- Small cost effective sizes.
- Available in low profile (0.33mm Max) for 0402.
- Applications: PA module, DC/DC converters, high frequency circuits, Bluetooth® modules, and portable electronics.
- 0201 X5R up to 0.22uF @ 6.3Vdc.
- 0402 X5R up to 2.2uF @ 4Vdc.

Wonder why your cell-phone got so small over the years? Thanks to Murata's dedication to technological advancements and dielectric research, we have engineered our line up of ultra small products for Big Solutions in tiny places... such as inside your trusted, cannot-live-without PDA and other handheld devices.

w w w . m u r a t a . c o m

CONTENTS
VOLUME 11 • ISSUE NUMBER 2

MARCH/APRIL 2009

2009 Buyers Guide and Resource Directory
HOW THE INFORMATION IS ORGANIZED:

The products and materials featured in this publication are intended to provide an overview of and reference for the passive component industry. *The Passive Component Industry eMagazine Buyers Guide and Resource Directory* is divided into four major categories; Capacitors, Resistors, Inductors, and Raw Materials. Each major section is divided into sub-categories by type of products/materials supported. Companies are listed alphabetically in each sub-category.

If you have updated information to submit, please e-mail us at info@paumanokgroup.com.

DEPARTMENTS

4 Letter from the Publisher
2009 Buyer's Guide Edition

6 Letter from the ECA
Counterfeiting: Beyond the Rolex

CAPACITORS

8 Ceramic Capacitor Suppliers

14 Tantalum Capacitor Suppliers

17 Aluminum Electrolytic Capacitor Suppliers

22 DC Film Capacitor Suppliers

25 AC Film Capacitor Suppliers

CAPACITORS

27 EDLC Supercapacitor Suppliers

29 Niobium Capacitor Suppliers

RESISTORS

30 Chip Resistor & Array

34 Resistor Network Suppliers
SIPs, DIPs, & IPD

37 Through-Hole Resistor Suppliers

INDUCTORS

39 Chip Inductor, Ferrite Bead, & Array

43 Ferrite Core Suppliers

RAW MATERIALS

44 Raw Material Suppliers to the Ceramic Capacitor Industry

48 Raw Material Suppliers to the Tantalum Capacitor Industry

51 Raw Material Suppliers to the Aluminum Electrolytic Capacitor Industry

53 Raw Material Suppliers to the Film Capacitor Industry

55 Raw Material Suppliers to the EDLC Supercapacitor Industry

55 Raw Material Suppliers of Niobium

56 Raw Material Suppliers to the Resistor Industry

Passive Component Industry is owned, published, and distributed by Paumanok Publications, Inc. Copyright 2000-2009 Paumanok Publications, Inc. All rights reserved. Parties submitting materials to Paumanok Publications represent and warrant that the submission does not violate any other party's proprietary rights. Paumanok Publications is not responsible for content. Article content, product names, services, and logos may be copyrighted, registered, and/or trademarked to their respective authors and/or companies.

LETTER FROM THE PUBLISHER

2009 Buyers Guide Directory

Welcome to the 2009 Passive Component Industry eMagazine Buyer's Guide edition. This directory is designed to show the link between component manufacturing and the raw materials consumed in their production. We hope you find the listings useful.

The focus of this directory is to list the global manufacturers of the core passive component product lines of capacitors, linear resistors, and inductors. The last section of the Buyer's Guide includes the key raw materials consumed in their manufacture—ceramic dielectric materials; electrode and termination materials; etched anode and cathode foils; tantalum ore, powder, and wire; metallized polyester and polypropylene film; resistor cores and substrates; and resistive inks.

For 2008, the total global market for capacitors was approximately $18 billion USD. The ceramic capacitors worldwide market was worth about $8 billion; thus, this market attracts the most competitors. Aluminum capacitors accounted for about $4 billion in worldwide revenues. Paper and plastic capacitors accounted for about $2.5 billion, of which DC film capacitors accounted for about $1 billion, and AC film capacitors about $1.5 billion worldwide. Tantalum capacitors accounted for approximately $2 billion in revenues.

The entire linear resistor business worldwide accounted for about $2.8 billion in revenues, while chip, bead, and ferrite core markets accounted for approximately $3.2 billion in 2008 revenues worldwide.

Thus, the combined market for the core passive components (capacitors, linear resistors, and inductors) in CY 2008 was about $24 billion. The 2009 CY forecast is for the market to be down about 20% in revenues, which would constitute a year-over-year loss of almost $5 billion in revenues.

For raw materials, a good rule of thumb is that primary raw materials account for approximately 25% to 30% of the component revenue by product type. Raw material supply markets to the global passive component industry were worth about $6.7 billion for 2008 (CY ending December). It is estimated these markets will decline by 20% as well in 2009 to about $5.4 billion in sales.

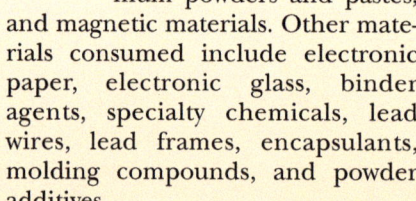

Dennis M. Zogbi

The primary raw materials consumed in the global passive components industry include tantalum powders, aluminum foils, polyester and polypropylene plastic films, ceramic dielectric materials, nickel powders, copper powders, palladium powders, and silver powders; as well as alumina substrates, ruthenium powders and pastes, and magnetic materials. Other materials consumed include electronic paper, electronic glass, binder agents, specialty chemicals, lead wires, lead frames, encapsulants, molding compounds, and powder additives.

We hope the directory is useful to you. Any feedback on how it might be updated or improved will be seriously considered. So please address your comments or suggestions regarding this directory to me directly at dennis@paumanokgroup.com.

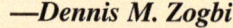
—*Dennis M. Zogbi*
Publisher, *Passive Component Industry*
President, Paumanok Publications, Inc.

PASSIVE COMPONENT INDUSTRY

Publisher
DENNIS M. ZOGBI

Director of Advertising
MITCHELL DEMSKO

Art Director
AMY DEMSKO

Editor
MELINDA ALLEN

Research Editor
NAUREEN S. HUQ

Advisory Board

Glyndwr Smith
Vishay Intertechnology, Inc.

Ian Clelland
***ITW* Paktron**

Pat Wastal
Avnet

Jim Wilson
MRA Laboratories

Bob Gourdeau
Vishay BCcomponents

Bob Willis
ECA President

Editorial and Advertising Office
224 High House Road, Suite 210
Cary, North Carolina 27513
(919) 468-0384   (919) 468-0386 Fax
www.paumanokgroup.com

The Electronic Components Association (ECA) represents the electronic components industry and the technologies, materials and supply chain associated with it. ECA offers market research, conferences, standards development under the EIA brand, issue advocacy, technology intelligence and collaborative efforts that help improve the business and technical expertise of its members.
For more information:
EIA Standards and Technology – (703) 907-8023, rjustus@ecaus.org
ECA Conferences and Expositions – (703) 907-8029, lrenzi@ecaus.org
ECA Membership/Market Research – (703) 907-8022, acappa@ecaus.org
ECA Programs and Services – (703) 907-8024, creid@ecaus.org
ECA Web Site Administration – (703) 907-8020, rtekin@ecaus.org
ECA members receive a 15% advertising discount for Passive Components Industry. Contact ECA at www.ecaus.org.

THE PAUMANOK GROUP

Get the latest in passive component market research from Paumanok Publications

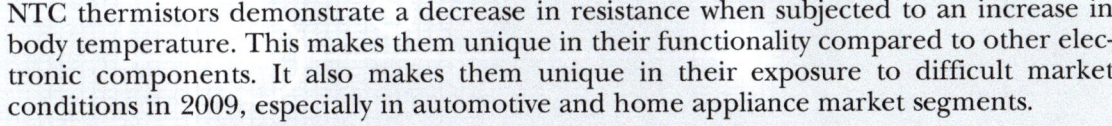

NTC Thermistors: World Market Outlook: 2009–2013

NTC thermistors demonstrate a decrease in resistance when subjected to an increase in body temperature. This makes them unique in their functionality compared to other electronic components. It also makes them unique in their exposure to difficult market conditions in 2009, especially in automotive and home appliance market segments.

This study addresses the global market for NTC thermistors by type, configuration, application, world region, competition, forecasts for auto consumption, forecasts for home appliance consumption, and forecasts for NTC consumption in digital electronics. NTC, as a sensor, has exposure to auto and home appliance segments that are dropping off rapidly.

Price: $1250 USD | **Additional Copies:** $125 USD
Downloadable (PDF) Copy: $1250 USD
Product Code: TC09
Web: www.paumanokgroup.com/market_reports/ppf/c/4/reports.asp

Passive Component Analysis of Quad Core Microprocessors: 2008: An Analysis and Comparison of AMD Phenom 9500 and Intel Core 2 Quad Q6600: Capacitors and Resistors

This is a teardown report that covers the analysis of two multi-core computer microprocessors—the AMD Phenom 9500 and the Intel Core 2 Quad Q6600—with an emphasis on the capacitor and resistor solutions that make them operate efficiently and effectively. The report includes a comparative analysis of capacitor and resistor content, including unit counts, component types, case sizes, performance values, and voltage ratings.

Price: $$2500 USD | **Additional Copies:** $$300 USD
Downloadable (PDF) Copy: $2500 USD
Product Code: Tear1
Web: www.paumanokgroup.com/market_reports/ppf/c/1/reports.asp

Ceramic Dielectric Materials: World Markets, Technologies & Opportunities: 2008–2013

The study looks at the world market for barium titanate and its high-purity precursors—barium carbonate and titanium dioxide. It also looks at both the captive and merchant markets for barium titanate based upon hydrothermal, oxalate, and sol-gel processes. There is also a look at the market for X7R, X5R, COG, and Y5V formulations. You will find pricing for materials, market shares in formulations, and barium titanate. The study looks closely at consumption by customer in the MLCC and PTC segments.

Price: $2500 USD | **Additional Copies:** $250 USD
Downloadable (PDF) Copy: $2500 USD | **Additional License:** $250 USD
Product Code: CeDiMat08
Web: www.paumanokgroup.com/market_reports/ppf/c/5/reports.asp

For more details on these reports and others, visit:
www.paumanokgroup.com

LETTER FROM ECA

Counterfeiting: Beyond the Rolex

By Bob Willis
President, ECA

When I was traveling on business in the early 1980s, I assumed the airlines and their employees were doing well. At least it appeared that way, as every flight attendant I met had a Rolex. In fact, I asked an acquaintance if she could get me one. No problem. That was until the crown dislodged and covered up the '6' on the watch's dial. That watch was worth what I paid for it, because it was obviously counterfeit.

At that time, I figured no harm, no foul. I was out 40 bucks but it was my own fault. I didn't realize the market impact of my $40 multiplied by a few million. Also, there was no consequential impact due to safety or performance problems. Jump ahead 20 years or so and it's a whole new ball of wax.

From Local to Global

Counterfeiting used to be primarily a local problem. It was harder to get away with when the customer was in close geographical proximity. Today's global marketplace and digital technologies make counterfeiting easier, as the counterfeiter is anonymous to the eventual buyer. As the New Yorker cartoon goes, "On the Internet, nobody knows you're a dog." Or, they don't know you're not a legitimate business person.

Counterfeiting today is a multi-billion dollar enterprise, mostly affecting brand names and consumer goods. The cases that get the most attention are fashion accessories, like the fake Rolex or the Gucci handbag that goes for $100 on a street corner in New York City. When you buy one of these, it might seem like a victimless exchange—both buyer and seller are complicit in the fact the item is a fake. But, like insider trading, there is a victim—the American economy, which loses approximately $200 billion a year from these illegal transactions.

In the electronics industry, the cases that get the most attention are identical copies of name-brand consumer products that offer sub-par performance at a reduced cost. This is an institution within itself, and for the most part, buyers know what they are getting.

A National Security Problem

The counterfeiting that concerns the electronic components industry is substituting counterfeit parts into name-brand products that are marketed as the real thing. This problem has impact for all parties in the manufacturing process and can result in performance and safety problems. After all, these products could be found in cars, airplanes, and defense and security systems. That represents a major security issue for the federal government.

The Department of Commerce (DoC) and the Naval Air Systems Command (NAVAIR) have teamed up to address the problem. The first major step is a study titled, "Counterfeits and the U.S. Industrial Base." For NAVAIR, counterfeit parts in the electronic supply chain have compromised the capabilities and readiness of DoD systems. This is viewed in parallel with the diminishing manufacturing and technological capabilities of the U.S. industrial base.

DoC's Office of Technology Evaluations (OTE) has conducted more than 50 industry studies and 125 surveys to monitor trends and benchmark industry performance. The OTE asked the ECA engineering program to assist with the industry survey to address counterfeits in the electronic components supply chain. Five separate, but related, surveys were sent to approximately 500 participants, including microchip and discrete electronic manufacturers, electronic board producers/assemblers, distributors, brokers, prime contractors, and DoD arsenals.

First Results at CARTS USA 2009

A complete review of these surveys will be presented at ECA's CARTS USA on March 30, 2009 in Jacksonville, Florida. A draft report is expected to be released in mid-2009 with the goal for industry and DoD to develop and implement best practices to ensure the integrity of the electronics supply chain. CARTS attendees will get an advance look at the results and receive information on best practices to mitigate supply chain risk.

While there are legal responsibilities for all parties to comply with federal regulations, more than 60% of survey respondents did not know which authorities to contact when they encounter counterfeits. For the most part, companies responded with two common approaches: Don't buy from China and be wary of brokers. In all cases, it takes two parties for counterfeiting to succeed—a seller and a willing customer. So now maybe more than ever, the bywords are "Buyer Beware."

If you want to find out in advance of the report's release about the impact of counterfeits on supply chain integrity, critical infrastructure and industrial capabilities, and what you can do about it, attend CARTS USA 2009. See www.ec-central.org for details.

ECA will continue to provide information about best practices and participate in legislative work to help manufacturers understand and combat this critical industry issue. ❑

Passive Component Industry eMagazine

Your global partner in the passive component marketplace

PCI eMagazine is always just a click away—we're now online!

- Search by keyword
- Launch to advertisers' Web sites and e-mail addresses
- Save issues to your computer to read later
- Access back issues

www.paumanokgroup.com/pci_magazine

Bookmark this link now!

Passive Component Industry eMagazine is read by key players in all aspects of the passive industry—from raw materials, to manufacturers, to end-users.

For more information contact Mitch Demsko at
<mitch@paumanokgroup.com>

CAPACITORS

2009 Buyers Guide and Resource Directory

Ceramic Capacitor Suppliers

AMERICAN TECHNICAL CERAMICS CORP. (AVX)
One Norden Lane
Huntington Station, NY 11746
(631) 622-4700
(631) 622-4748 (Fax)
Products: Specialty Ceramic Capacitors (MLC & SLC).

Key Plant Location:

AMERICAN TECHNICAL CERAMICS CORP. (AVX)
2201 Corporate Square Blvd.
Jacksonville, FL 32216
(904) 726-3426
(904) 725-2279 (Fax)
Products: Single Layered Ceramic High Voltage and Microwave Ceramic Capacitors. A high-end military and Hi-REL commercial supplier of advanced single and multi-layered ceramic capacitors for high voltage and high-frequency applications. Florida operations also produce thin- film and LTCC ceramics.

ANAREN CERAMICS, INC.
27 Northwestern Drive
Salem, NH 03079
(603) 898-2883
(603) 898-4273 (Fax)
Products: Ceramic Substrates, LTCC Materials, Resistors, Terminations, Attenuators, Couplers and Custom Products.

AVX CORP.
PO Box 867
Myrtle Beach, SC 29578
(843) 448-9411
Products: MLCC and MLC Leaded.
AVX is a publicly traded company on the NYSE whose majority shareholder is Kyocera Corporation of Japan. AVX is one of the largest producers of ceramic capacitors in the world, including multi-layered, single layered and specialty ceramic capacitors– innovative products like the LICA chip, and value-added and application specific components for high voltage and high frequency.

Key plant locations are as follows:

AVX COLERAINE
5 Hillmans Way
Coleraine, BT52 2DA
Londonderry
United Kingdom
Products: MLCC and MLC Leaded.

AVX SAN SALVADOR
Calle Cojutepeque #4-2
Zona Franca San Bartolo
Ilopango San Salvador
Products: MLC Leaded.

AVX MEXICO
Avio Excelente S de R.L. de C.V.
Ave. San Lorenzo No. 651
Area Riberena 32310
Cd Juarez Chihuahua
Mexico
Products: MLC Leaded.

AVX/KYOCERA SINGAPORE
151 Lor Chuan
#06-05/06 New Tech Park
Singapore 556741
Products: MLCC and MLC Leaded.

AVX CORP.
OLEAN ADVANCED PRODUCTS
1695 Seneca Avenue
PO Box 493
Olean, NY 14760
(716) 372-6611
(716) 372-6635 (Fax)
Products: Specialty Ceramics.
Additional ceramic related plants in Colorado Springs, CO, Raleigh, NC, and Israel. New plant construction in China.

AVX-THOMSON-LCC
Av. Du Colonel Prat.
TPC 21850 St. Appollinaire
France
(33-8) 071-7400
Products: High Voltage Ceramic Capacitors.

AVX CORP.

Product: MLCC (including FLEXISAFE™ & FLEXITERM®)

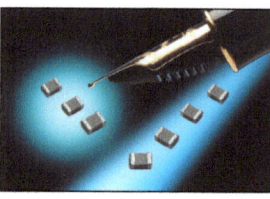

AVX offers a broad range of MLCC including NP0, X7R, X5R, X8R and Y5V dielectrics. Case size options 0201, 0402, 0603, 0805, 1206, 1210, 1812, 2220, and even much larger sizes and custom configurations are available.
Recent developments include a family of industry leading, ultra low inductance devices (LGA), which are designed for high data rate decoupling applications. They offer superior inductance performance, compared to the currently used IDC (inter-digitated capacitor) design.
The award winning, FLEXISAFE™ capacitor is another technology first. It has a unique combination of an internal cascade design and shock absorbing, FLEXITERM® termination technology for safety critical and unlimited current applications.
Another new product is our LT series that has a profile height of just 0.35mm for an 0402, 1uF, 4V X5R device. A family of parts from 0201 to 0805 is available.

Tim Hollander
Marketing Manager – Ceramic
thollander@avxus.com
Phone: (843) 946-0524

CAPACITORS

DIELECTRIC LABS (DOVER CORP.)
2777 Route 20 East
Cazenovia, NY 13035
(315) 655-8710
(315) 655-0445 (Fax)
Products: Specialty High Frequency Ceramic Capacitors.

EPCOS AG
(EPCOS and TDK will merge in 2009)
P.O. Box 80 17 09
81617 Munich
Germany
(49) 896-3609
(49) 896-362-2689 (Fax)
Products: Multi-layer serial ceramic capacitors (MLSC).

EUROFARAD
93 Rue Oberkampf
F-75540 Paris Cedex 11
France
(33-14) 923-1000
Products: Specialty MLCC for Defense and Aerospace.

FENGHUA ADVANCED TECHNOLOGY CO., LTD.
18th Fenghua Road
Zhaoqing City, Guangdong Province
China
(86-758) 286-5325
(86-758) 286-5174 (Fax)
Products: Capacitors; Aluminum Electrolytic and Ceramic.

GREATBATCH, INC.
10000 Wehrle Drive
Clarence, NY 14031
(716) 759-5600
(716) 759-5560 (Fax)
Products: Ceramic feed-through, as well as high voltage ceramic and specialty ceramic for downhole pump.

HOLY STONE ENTERPRISE CO., LTD.

Products: Multilayer Ceramic Capacitors, Leaded Disc Capacitors

Holy Stone Enterprise Company is a Taiwan-based manufacturer of ceramic-based components that are designed for customer manufacturability and performance.

Holy Stone offers a broad line of ceramic capacitors, but specializes in:
- Safety Certified SMT Capacitors, X1/Y2, X2/Y3
- High Voltage MLCC's for power applications, voltage multipliers, invertors, ballasts, etc.
- High Capacitance NPO for XDSL applications
- Trigger Capacitors for DSC strobe circuits
- Low Distortion X7E Capacitors for oscillation and filter circuits
- X6R/X7S Capacitors where operating temperatures are too high for X5R
- "SuperTerm" polymer enhanced terminations for board flexure durability
- Proprietary coatings to enhance cleaning properties during assembly
- Safety Certified X1/Y1, X1/Y2, X2/Y3 Leaded Ceramic Disc Capacitors

Holy Stone Capacitors are produced in modern facilities certified to ISO-14000, ISO-9001 and QS-9000. Holy Stone products are fully RoHS compliant.

In America:
HolyStone International
Steve Jouflas, National Sales Director
26395 Jefferson Ave., Suite H
Murrieta, CA 92562
Phone: 951-696-4300
info@holystonecaps.com
www.holystonecaps.com

In Europe:
HolyStone (Europe) Limited
sales@holystonecaps.co.uk
www.holystoneeurope.com

In Asia:
HolyStone Enterprise Co., Ltd.
Inquiry@holystone.com.tw
www.holystone.com.tw

JOHANSON DIELECTRICS, INC.
15191 Bledsoe Street
Sylmar, CA 91342
(818) 364-9800
(818) 364-6100 (Fax)
Products: MLCC, SLC, Specialty and LTCC.

KEKO VARICON
Grajski trg 15
SI-8360 Zuzemberk
Slovenia
(38-67) 388-5178
(38-67) 388-5158 (Fax)
Products: Ceramic Capacitors.

KEMET ELECTRONICS

Products: High CV, Commercial High Voltage and Open-Mode Ceramic Capacitors

Kemet Electronics Corporation continues to broaden its ceramic portfolio with the introduction of its newest high capacitance

ceramic components, designed to provide circuit functions like decoupling and output filtering for the computer, graphic, power supply, satellite, automotive, and industrial market segments. Extended CV (capacitance/unit volume) surface mount offerings are available in X5R and X7R dielectrics for 0603 through 1206 case sizes. Kemet's latest commercial high voltage surface-mount capacitors are designed for applications demanding reliability at high voltage; including rated voltages from 500 to 3000V in C0G and X7R dielectrics, and are available in EIA case sizes from 0805 to 2225. Newly released 1812 case open-mode capacitors are designed to reduce low insulation resistance and short circuit conditions in board flex situations. Potential targets include automotive and power supply customers. All parts are environmentally friendly, in compliance with RoHS legislation.

Johnny Boan
capmaster@kemet.com
Kemet Corp.
2835 Kemet Way
Simpsonville, SC 29681
Phone: (864) 963-6300
www.kemet.com

CAPACITORS

KYOCERA CORP.
6 Takeda Tobadono-cho
Fushimi-ku, Kyoto 612-8501
Japan
(81-75) 604-3500
Products: MLCC; Primarily for Japan Wireless Markets.
Majority owner of AVX Corporation.

MARUWA CO., LTD.
3-83, Minamihonjigahara-cho
Owariasahi-city, Aichi-pref., 488-0044
Japan
(81-56) 151-0841
(81-56) 151-0845 (Fax)
Products: Fixal Ceramic Disc, Temperature Compensated, Multi-Layer Chip, Semiconductor Type and Feed-Through.

MURATA MANUFACTURING CO., LTD.
10-1, Higashikotari 1-chome
Nagaokakyo-shi, Kyoto 617-8555
Japan
(81-75) 951-9111
Products: MLCC.
One of the world's largest producer of ceramic chip capacitors.

MURATA ELECTRONICS, N.A., INC.

Product: High-Cap MLCC

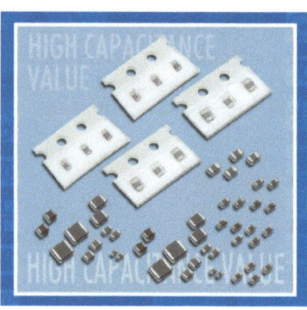

Murata's Technology supports the electronic equipment manufacturers by offering industry leading quality components. Why use Murata's High-Cap MLCC ceramic chips over Ta and Al capacitors?
Murata's ceramic chips offer better reliability, electrical performance, no polarity, and allow engineers to downsize designs with various package options from 0402 X5R 4.7 µF to 1206 X5R 100 µF. The GRM product series is offered with a Nickel (Ni) barrier termination plated with Matte Tin (Sn), and is RoHS compliant.
The Murata Hi-Cap products are available in X5R, X7R, and other TC's with a capacitance of 1µF and larger. This line of high volumetric capacitance is available in 4 V, 6.3 V, 10 V, 16 V, 2 V, and 50 V that has excellent high frequency noise absorption performance compared to Ta and Al caps. Stringent dimensional tolerances allow highly reliable, high-speed automatic chip placement on PCBs. For bypass/decoupling applications above 10 kHz, the impedance of MLCC is much lower than Ta capacitors. Therefore, MLCC capacitors with values of half to one-fifth the value of the Ta-cap can be used instead.
Consider Murata's Hi-Cap as your Ta capacitor replacement.

Mark Waugh
Sr. Product Manager
mwaugh@murata.com
2200 Lake Park Drive
Smyrna, GA 30080
Phone: (770) 436-1300
www.murata.com

NIC COMPONENTS CORP.

Product: Ceramic Capacitors

NIC Components Corp. is a leading supplier of passive components, including a broad line of SMT and leaded ceramic capacitors. SMT MLCCs are available in capacitance values from 0.5pF to 100uF and in case sizes range from 0201 to 2225. NPO, X7R, X5R and Y5V temperature coefficients cover voltage ratings from 4VDC to 5000VDC. SMT ceramic chip X1/Y2 & X2/Y3 safety capacitors (UL and TUV safety agency certified) are available from NIC. Additional ceramic capacitor products include: SMT capacitor arrays, radial and axial mono capacitors and high voltage radial leaded ceramic disc capacitors with voltage ratings to 15,000VDC. NIC has a direct presence in New York, California, Florida, Puerto Rico and Canada, including East and West Coast warehouses. Additional NIC European (NIC Components Europe) and South-East Asian (NIC Components Asia) divisions provide sales, product logistics and technical support to NIC customers in these regions. FREE "Design Your Own" Product Sample Kits at www.niccomp.com.

Eric Moller
National Sales Manager
sales@niccomp.com
70 Maxess Road
Melville, NY 11747 USA
Phone: (631) 396-7500
www.niccomp.com

NOVACAP (DOVER CORP.)
25111 Anza Drive
Valencia, CA 91355
(661) 295-5920
Products: MLCCs, capacitor arrays, radial leaded capacitors, and pulsed power capacitors for detonation circuitry.

OXLEY DEVELOPMENTS CO.
Unit 1, Furness Business Park
Peter Green Way, Barrow-in-Furness. LA14 2PE
UK
(44-0-122) 984-0600
(44-0-122) 943-1342 (Fax)
Products: Specialty Ceramics.

CAPACITORS

PANASONIC ELECTRONIC DEVICES CO., LTD.
1006 Oaza Kadoma
Kadama City
Osaka 571-8506
Japan
(81-66) 906-1652
Products: MLCCs, MLC Leaded and SLCs.
Panasonic is primarily known as a blanket supplier of both fixed and variable on-board capacitors; they supply as much product by type, configuration and dielectric as they possibly can on a worldwide basis. Panasonic has been widely successful in DC film capacitors and aluminum electrolytic capacitors outside of Japan. They also produce ceramic capacitors and tantalum capacitors for captive consumption and merchant market sales.

PRESIDIO COMPONENTS, INC.
7169 Construction Court
San Diego, CA 92121
(858) 578-9390
(858) 578-6225 (Fax)
Products: High Voltage MLC and SLC.

PROSPERITY DIELECTRIC CO. (WALSIN)
10F, No. 480, Rueiguang Rd.
Neihu Chiu 114, Taipei
Taiwan
(886-22) 797-6698
(886-22) 797-1766 (Fax)
Products: MLCC, Chip-R, Ceramic Dielectric Powders, Disc-Type Semi-Conductive Capacitor Elements.

RCD COMPONENTS INC., RESISTORS-CAPACITORS-COILS-DELAY LINES

Product: Ceramic Capacitor Suppliers

RESISTORS • CAPACITORS • COILS • DELAY LINES

RCD offers one of the industry's widest range of passives, with many new models. Worldwide manufacturing capability ensures lowest pricing from smallest to highest volumes. Two billion piece inventory assures immediate delivery. RCD's unique SWIFT™ department can supply almost any tantalum or ceramic capacitor, including Mil-spec items & custom R, C, R/C networks within 2 weeks. Zero defects guaranteed! Custom products… our specialty for 40 years!

- MLCC chips from 0201 to 3035, 0.47pF - 100uF
- Axial- & radial-lead Ceragold™ ceramic construction
- COG (NPO), X5R, X7R, Z5U & Y5V
- Tantalum chip Tangold™ caps, 0.1uF – 470uF
- Radial-lead tantalum, 0.047uF – 680uF
- Specialty models: low ESR, high-frequency, high-voltage (30KV)
- Precision tolerances to ±1%, matching to 0.1%
- Arrays, networks, hybrids, IPC's
- Shunts and current sensors to 15KA
- Resistors: film/foil/WW/composition to 300KV
- Shield & non-shielded inductors/transformers
- Active/passive/programmable delay lines

Mary Jo Allen
Sales Manager
sales@rcdcomponents.com
520 East Industrial Park Drive
Manchester, NH 03109
Phone: (603) 669-0054
www.rcdcomponents.com

SAMSUNG ELECTRO MECHANICS CO.
314 Maetan-3-dong
YeongTong-gu, Suwon City
Gyeonggi Province 442-743
Korea
(82-31) 210-5114
(82-31) 210-6363 (Fax)
Products: MLCC.
Samsung (SEMCO) has long been considered the Korean company that would make the first leap outside of the Asian perimeter and into world MLCC markets. The company has emerged as a major MLCC producer on a global scale.

Other Location:

SAMSUNG ELECTRO MECHANICS CO., LTD.
Block 5, Calamba Premiere International Park
Barangay Batino, Calamba City
Laguna, Philippines
(63-49) 545-6001
(63-49) 545-2348 (Fax)

SAMWHA CAPACITOR CO., LTD.
Samyoung Bldg.
587-8 Sinsa-dong
Gangnam-gu, Seoul
Korea
(02) 545-5600
(02) 518-8501 (Fax)
Products: MLCC, Chip Varistors, Chip NTC.

SPECTRUM CONTROL
8061 Avonia Road
Fairview, PA 16415
(814) 474-1571
(814) 474-3110 (Fax)
Products: Tubular Ceramic Capacitors, Switch Mode Power Supplies, Discoidal Capacitor, Specialty Medical Capacitors.

SYFER TECHNOLOGY LIMITED (DOVER CORP.)
Old Stoke Rd.
Arminghall, Norwich, Norfolk
NR14 8SQ
United Kingdom
(44-160) 372-3310
Products: Ceramic Multi-Layer Capacitors.

TAIYO YUDEN CO. LTD
6-16-20 Ueno, Taito-ku,
Tokyo 110-0005
Japan
(81-33) 832-0101
(81-33) 832-0105 (Fax)

CAPACITORS

TAIYO YUDEN (U.S.A), INC.

Product: 4.7μF MLCC in 0402 Case Size

Taiyo Yuden (U.S.A.) Inc., offers a complete line of high-performance multilayer ceramic capacitors (MLCC), the latest of which is the AMK105BJ475MV for decoupling power-line circuits in cell phones, digital still cameras and other battery-powered portable electronics. The AMK105BJ is the first MLCC to achieve a capacitance rating of 4.7μF in the 1.0 x 0.5 x 0.5mm (EIA 0402) case size, allowing designers to obtain the desired capacitance value from a single 0402-size device, where two 2.2μF-rated 0402 capacitors were previously required. This effectively doubles available capacitance for the same component mounting area, saving cost and simplifying manufacturability by reducing parts count by 50 percent. The X5R-rated AMK105BJ475MV provides a 4.7μF (± 20%) capacitance range, 4 VDC voltage rating and an operating temperature range of -55 to +85 C. For further information, visit the Taiyo Yuden (U.S.A.) Inc. web site at www.yuden.us.

Yaeko Minamikawa
Marketing Analyst
sales@t-yden.com
1930 Thoreau Drive, Ste. 190
Schaumburg, IL 60173
Phone: (847) 925-0888
www.yuden.us

TDK CORP.
(TDK and EPCOS will merge in 2009)
1-13-1, Nihonbashi,
Chuo-ku, Tokyo 103-8272
Japan
(88-33) 278-5111
Products: MLCC.
TDK produces large volumes of MLCC in Japan and terminates them globally.

TEAM YOUNG ADVANCED CERAMICS
No 15 Kung 3 Road
Ping Chen Industrial Zone Taoyuan
Taiwan, R.O.C.
(886-03) 469-9559
(886-03) 469-9381 (Fax)
Products: Bulk MLCC.

TEMEX CERAMICS
Formerly Tekelec Components
Parc Industriel Bersol
Voie Romaine
33600 Pessac I
France
(33-055) 646-6666
(33-055) 636-3198 (Fax)
Products: High-Frequency Ceramic Capacitors.

TRIGON COMPONENTS, INC.
939 Mariner Street
Brea, CA 92821
(714) 990-1367
Products: Passive and electromagnetic components.

TRG COMPONENTS DIRECT (REPUBLIC ELECTRONICS CORP.)
5801 Lee Highway
Arlington, VA 22207
(703) 584-8545
(703) 533-2079 (Fax)
Products: Specialty Ceramic Capacitors.

TRS TECHNOLOGIES, INC.
2820 E. College Avenue, Suite J
State College, PA 16801
(814) 238-7485
(814) 238-7539 (Fax)
Products: Specialty Ceramic Capacitors.

TUSONIX, INC.
7741 N. Business Park Drive
Tucson, AZ 85743
(520) 744-0400
(520) 744-6155 (Fax)
Products: Tubular Ceramics, Disc Capacitors.

TTI, INC.

Product: Never Short on Solutions

Headquartered in Fort Worth, Texas, TTI, Inc. is a distributor specialist of passive, interconnect, electromechanical, and discrete components. TTI is the distributor of choice for industrial and consumer electronic manufacturers worldwide.

TTI's product line includes: resistors, capacitors, connectors, potentiometers, trimmers, magnetic and circuit protection components, wire and cable, wire management, identification products, application tools, electromechanical devices, and discrete components. We distribute these products from a broad line of manufacturers.

TTI strives to be the industry's preferred information source by offering, through the ttiMarketEye blog, the latest IP&E technology and market information, technical seminars, RoHS seminars, industry research reports and much more.

TTI employs more than 2,000 people at more than 50 locations throughout North America, Europe, and Asia.

Sales Contact: information@ttiinc.com
2441 Northeast Pkwy
Fort Worth, Texas 76106
Phone: 800-CALL-TTI (225-5884)
www.ttiinc.com

CAPACITORS

UNION TECHNOLOGY CORPORATION

Product: Capacitors – Ceramic

UTC offers specialty ceramic components.
- SMPS Capacitors (MIL-PRF 49470)
- High Voltage Radials (DSCC Qualified)
- SMT High Voltage Chips
- Safety Certified Ceramic Chips
- SMT Multi-Layer Ceramic Chips
- Large Body Size SMT Power Chips
- Feed-Thru Discoidal Capacitors
- Multi-Pin Planar Array Capacitors

Our manufacturing facility is an ISO 9000-2001 certified plant, as well as being MIL-790 approved.
UTC is also certified by the Small Business Administration, as a: SDB-"8a"

Tom Fatica
Vice President of Sales & Marketing
info@uniontechcorp.com
718 Monterey Pass Rd.
Monterey Park, CA 91754
Phone: (323) 266-6603
www.uniontechcorp.com

WALSIN TECHNOLOGY CORPORATION

Product: Multi-Layer Ceramic Capacitors

Walsin Technology Corporation offers a full range of SMT, ROHS compliance multi-layer ceramic capacitors, covering commodity to high CV, precision capacitors in Y5V, NP0, X7R, X5R & X6S dielectrics of all standard sizes of 0201, 0402, 0603, 0805, 1206, 1210, 1812 & 2211. Specialty like arrays, low inductance (with X7R dielectric and 0612 in size), high-Q/low-VESR capacitors are offered for noise suppression in high-speed processors & high frequency applications. High voltage ranges 200V to 5KV are commonly used industrial applications such as power management, inverter and lighting ballast. Walsin also offers X2Y3, X1Y2 category SMD safety certificated capacitors (with NPO & X7R dielectrics and 1808, 1812 & 2211 three standard sizes) for AC circuit protection.

HQ, Taiwan - Charles Chu
charleschu@passivecomponent.com
Phone: +886-3-475-8711

North America - Lee Ku
leeku@walsin-usa.com
Phone: +1-510-713-1190

Europe, Germany - Alphonse
aw@passivecomponent.com
Phone: 49-0-89-9308-6475
www.passivecomponent.com

VISHAY INTERTECHNOLOGY, INC.
63 Lancaster Avenue
Malvern, PA 19355-2143
(610) 644-1300
(610) 296-0657 (Fax)
Products: Tantalum Capacitors (Molded Chip Tantalum Capacitors, Coated Chip Tantalum Capacitors, Leadframeless Chip Tantalum Capacitors, Through-Hole Solid Tantalum Capacitors, Wet Tantalum Capacitors, Capacitor Arrays and Modules); Ceramic Capacitors (Multilayer Chip Capacitors, Disc Capacitors); Film Capacitors; Power Capacitors; Heavy-Current Capacitors; Aluminum Capacitors; Silicon RF Capacitors.
Capacitor Brands:
Vishay BCcomponents
Vishay Cera-Mite
Vishay Draloric
Vishay ESTA
Vishay Roederstein
Vishay Sprague (includes Mallory [NACC] and Tansitor)
Vishay Vitramon
and others

VITRAMON (VISHAY)
Vishay Electronics GmbH
Geheimrat-Rosenthal-Str. 100
95100 Selb
Germany
Products: MLCC for Automotive and Industrial markets.

WRIGHT CAPACITORS, INC.
2610 So. Oak Ave.
Santa Ana, CA 92707
(714) 546-2490
(714) 546-1709 (Fax)
Products: High Voltage Ceramic.

YAGEO CORP.
3F No. 233-1 Pao-Chiao Rd
Hsin Tien, Taipei
Taiwan
(88-622) 917-7555
(88-622) 917-3789 (Fax)

CAPACITORS

TTI, INC.

Product: Never Short on Solutions

Headquartered in Fort Worth, Texas, TTI, Inc. is a distributor specialist of passive, interconnect, electromechanical, and discrete components. TTI is the distributor of choice for industrial and consumer electronic manufacturers worldwide.

TTI's product line includes: resistors, capacitors, connectors, potentiometers, trimmers, magnetic and circuit protection components, wire and cable, wire management, identification products, application tools, electromechanical devices, and discrete components. We distribute these products from a broad line of manufacturers.

TTI strives to be the industry's preferred information source by offering, through the tti|MarketEye blog, the latest IP&E technology and market information, technical seminars, RoHS seminars, industry research reports and much more.

TTI employs more than 2,000 people at more than 50 locations throughout North America, Europe, and Asia.

Sales Contact: information@ttiinc.com
2441 Northeast Pkwy
Fort Worth, Texas 76106
Phone: 800-CALL-TTI (225-5884)
www.ttiinc.com

VISHAY INTERTECHNOLOGY, INC.
63 Lancaster Avenue
Malvern, PA 19355-2143
(610) 644-1300
(610) 296-0657 (Fax)
Products: Tantalum Capacitors (Molded Chip Tantalum Capacitors, Coated Chip Tantalum Capacitors, Leadframeless Chip Tantalum Capacitors, Through-Hole Solid Tantalum Capacitors, Wet Tantalum Capacitors, Capacitor Arrays and Modules); Ceramic Capacitors (Multilayer Chip Capacitors, Disc Capacitors); Film Capacitors; Power Capacitors; Heavy-Current Capacitors; Aluminum Capacitors; Silicon RF Capacitors.
  Capacitor Brands:
  Vishay BCcomponents
  Vishay Cera-Mite
  Vishay Draloric
  Vishay ESTA
  Vishay Roederstein
  Vishay Sprague (includes Mallory [NACC] and Tansitor)
  Vishay Vitramon
  and others

In the Next Issue
New Products, Markets & Opportunities

If interested in submitting an article, contact Mitch Demsko by April 30, 2009 at

mitch@paumanokgroup.com
or call (919) 601-0112

www.paumanokgroup.com

CAPACITORS

MATSUO ELECTRIC CO., LTD.
3-5-3, Sennaricho 3-chome
Toyonaka Shi, Osaka 561-8558
Japan
(06-6) 332-0871
(06-6) 331-1386 (Fax)
Products: Tantalum Capacitors Primarily for Asian Markets.

NEC TOKIN CORP.
Chiyoda First Bldg., 8-1, Nishi-Kanda 3-chome
Chiyoda-ku, Tokyo 101-8362
Japan
(81-33) 515-9222
(81-33) 515-9223 (Fax)

NICHICON CORP.
Karasumadori Oike-agaru Nakagyo-ku
Kyoto, 604-0845
Japan
(81-75) 231-8461
(81-75) 256-4158 (Fax)
Products: Aluminum and Tantalum Capacitors.

PALM BEACH COMPONENTS, INC.
5401 N. Haverhill Rd, Suite #101
West Palm Beach, FL 33407
(561) 478-7440
(561) 478-7470 (Fax)
Products: Capacitor Assemblies.

PANASONIC ELECTRONIC DEVICES CO., LTD.
1006 Oaza Kadoma
Kadama City
Osaka 571-8506
Japan
(81-66) 906-1652
Products: MLCCs, MLC Leaded and SLCs.
Panasonic is primarily known as a blanket supplier of both fixed and variable on-board capacitors; they supply as much product by type, configuration and dielectric as they possibly can on a worldwide basis. Panasonic has been widely successful in DC film capacitors and aluminum electrolytic capacitors outside of Japan. They also produce ceramic capacitors and tantalum capacitors for captive consumption and merchant market sales.

ROHM CO., LTD.
21 Saiin Mizosaki-cho
Ukyo-ku, Kyoto 615-8585
Japan
(81-75) 311-2121
(81-75) 315-0172 (Fax)
Products: Tantalum Chip Capacitors.

ROHM ELECTRONICS USA, LLC

Product: Tantalum Capacitors

ROHM offers a broad lineup of tantalum capacitors. The TCT series features a novel package structure that allows for a larger tantalum element than conventional units of the same size, resulting in greater capacitances, while the TCO series features low ESR characteristics, making them ideally suited for high-frequency noise absorption. Also available is the high-performance TCFG series with a failsafe design that prevents damage to peripheral components in the event of malfunction.
For more information on these products or to find a ROHM representative in your area, please visit www.rohmelectronics.com.

Marketing76@rohmelectronics.com
10145 Pacific Heights Blvd., Ste. 1000
San Diego, CA 92121
Phone: 888-775-ROHM
www.rohmelectronics.com

SAMSUNG ELECTRO MECHANICS CO., LTD.
Block 5, Calamba Premiere International Park
Barangay Batino, Calamba City
Laguna, Philippines
(63-49) 545-6001
(63-49) 545-2348 (Fax)
Products: A Small Line of Tantalum Capacitors.

SANYO ELECTRONIC DEVICE CO.
1-1 SANYO-cho, Daito, Osaka, 574-8534
Japan
(072) 872-0011
(072) 873-9858 (Fax)

TPC / HILTON CAPACITOR

Product: SWT Wet Tantalum Capacitors

Miniature silver cased, non-hermetic wet tantalum capacitors that cross with Vishay/Mallory MTP capacitor line. These capacitors have voltage ratings up to 100V and capacitance ratings up to 1000µF. Guaranteed DCL specs are lowest in the industry. Lead times are normally stock to 4 weeks.

Sales
Assistant Manager
Sales@tantalum-pellet.com
21421 N. 14th Avenue
Phoenix, AZ 85027
Phone: (623) 582-5555
www.tantalum-pellet.com

CAPACITORS

TTI, INC.

Product: Never Short on Solutions

Headquartered in Fort Worth, Texas, TTI, Inc. is a distributor specialist of passive, interconnect, electromechanical, and discrete components. TTI is the distributor of choice for industrial and consumer electronic manufacturers worldwide.

TTI's product line includes: resistors, capacitors, connectors, potentiometers, trimmers, magnetic and circuit protection components, wire and cable, wire management, identification products, application tools, electromechanical devices, and discrete components. We distribute these products from a broad line of manufacturers.

TTI strives to be the industry's preferred information source by offering, through the ttiMarketEye blog, the latest IP&E technology and market information, technical seminars, RoHS seminars, industry research reports and much more.

TTI employs more than 2,000 people at more than 50 locations throughout North America, Europe, and Asia.

Sales Contact: information@ttiinc.com
2441 Northeast Pkwy
Fort Worth, Texas 76106
Phone: 800-CALL-TTI (225-5884)
www.ttiinc.com

VISHAY INTERTECHNOLOGY, INC.
63 Lancaster Avenue
Malvern, PA 19355-2143
(610) 644-1300
(610) 296-0657 (Fax)
Products: Tantalum Capacitors (Molded Chip Tantalum Capacitors, Coated Chip Tantalum Capacitors, Leadframeless Chip Tantalum Capacitors, Through-Hole Solid Tantalum Capacitors, Wet Tantalum Capacitors, Capacitor Arrays and Modules); Ceramic Capacitors (Multilayer Chip Capacitors, Disc Capacitors); Film Capacitors; Power Capacitors; Heavy-Current Capacitors; Aluminum Capacitors; Silicon RF Capacitors.
Capacitor Brands:
Vishay BCcomponents
Vishay Cera-Mite
Vishay Draloric
Vishay ESTA
Vishay Roederstein
Vishay Sprague (includes Mallory [NACC] and Tansitor)
Vishay Vitramon
and others

In the Next Issue
New Products, Markets & Opportunities

If interested in submitting an article, contact Mitch Demsko by April 30, 2009 at
mitch@paumanokgroup.com
or call (919) 601-0112

www.paumanokgroup.com

CAPACITORS

Aluminum Electrolytic Capacitor Suppliers

CORNELL-DUBILIER ELECTRONICS CO.
140 Technology Place
Liberty, SC 29657
USA
(864) 843-2277
(864) 843-3800 (Fax)

DAEWOO ELECTRONIC COMPONENTS (PARTSNIC)
543 Dangjeong-dong, Kunpo-si
Kyoungki-do 435-030
Korea
(82-31) 428-1600
(82-31) 453-1563 (Fax)

DAE YEONG ELECTRONICS CO., LTD.
116 Songhak-ri
Gasan-myeon, Chilgok-gun
Gyeongsangbuk-do Province 718-911
Korea
(82-54) 975-8009~10
(82-54) 975-8011 (Fax)

EFC/WESCO

Products: Film & Electrolytic Capacitors

EFC/WESCO, your complete source for Film and Power capacitors, is proud to introduce their line of VDE and UL recognized EMI suppression capacitors and Capxon line of Electrolytics.

The EMI Suppression Films are "X" and "Y" class radial lead Polypropylene Box style with voltage ratings to 275 VAC in the "X2" class and capacitance to 3.3 µF.

Our Capxon line of Electrolytic capacitors offers all of the popular series including "Snap In" and "Surface Mount" packaging. Please contact the factory for our latest catalog.

EFC/WESCO is proud to be an ISO 9001:2000 certified company.

Bob Fountain
Sales Manager
fountain@filmcapacitors.com
41 Interstate Lane
Waterbury, CT 06705
Phone: 203-755-5629
www.filmcapacitors.com

ELNA CO., LTD.
3-8-11 Shin-Yokohama
Kouhoku-ku
Yokohama-city 222-0033
Kanagawa
Japan
(81-45) 470-7254
(81-45) 470-7260 (Fax)
Products: Aluminum Electrolytic Capacitors for Consumer Electronics Industry in Japan and a Line of Solid Tantalum Capacitors.

EPCOS, INC.

Product: Aluminum Electrolytic Capacitors

EPCOS offers a broad range of aluminum electrolytic capacitors, including radial, axial, solder-star, snap-in, solder-pin, and screw terminal capacitor types. EPCOS is the market leader for screw terminal capacitors, especially high voltage types for use in inverters and drives. These are used in appliances, transportation and traction, and automation and conveyance equipment. Voltage ratings of up to 500VDC are available, and capacitance values as high as 18,000µF. Ripple current ratings of 60A is available when optional bottom cooling thermo-pads are utilized.

New product families include types resistant to high temperature and high vibration for automotive applications. Hybrid and electric vehicles require bulk storage for applications including airbag, electronic steering, load-leveling suspension, crankshaft starter/alternator and electronic braking systems. Product range available includes 105, 125, and 150°C rated types as well as high resistance to vibration as high as 40g. Capacitance values of up to 20,000µF and voltage ratings up to 63V.

Matt Reynolds
Product Marketing Manager
matt.reynolds@epcos.com
186 Wood Avenue South
Iselin, NJ 08830
Phone: (800) 888-7729
www.epcos.com

EPCOS AG
(EPCOS and TDK will merge in 2009)
P.O. Box 80 17 09
81617 Munich
Germany
(49) 896-3609
(49) 896-362-2689 (Fax)
Products: Snap-in aluminum electrolytic capacitors specifically for link-circuit applications in frequency converters.

EUROFARAD
93 rue Oberkampf
F-75540 Paris Cedex 11
France
(33-14) 923-1000
(33-14) 357-0533 (Fax)

EVOX RIFA AB (KEMET)
Thörnblads väg 6
SE-386 90 Färjestaden
Sweden
(46-48) 556-3900

FUJITSU TOWA ELECTRON, LTD.
4-1-1 Kamikodanaka, Nakahara-ku
Kawasaki, Kanagawa 211-8588
Japan
(81-44) 777-1111

HITACHI AIC
South Shin-Otsuka 11F
2-25-15, Minami-Otsuka
Toshima-ku, Tokyo 170-0005
Japan
(81-35) 319-5581
(81-35) 319-5855 (Fax)

KEMET ELECTRONICS CORPORATION
2835 KEMET Way
Simpsonville, SC 29681
(864) 963-6300

CAPACITORS

JIANGHAI EUROPE GMBH

Product: Aluminum Electrolytic Capacitors

With more than 2.5 billion electrolytic capacitors per year, Jianghai is one of the largest manufacturers in China: SMD, radial, snap-in, and screw capacitors for all voltages 6 to 550 V. Quality certifications such as ISO 9001, ISO 14001, UL and QS 9000 stand for a high-reliable, quality production. Of course all components are lead-free and conform to WEEE/RoHS. Jianghai capacitors are used in all kinds of professional industry applications, especially power inverters, drives and control systems, UPS, and lighting ballast. Due to very successful joint ventures (e.g. Hitachi AIC) and partnerships to other leading Japanese capacitor manufacturers, Jianghai products are sold under their own and external brand labels successfully worldwide. As our focus is on high-end industry applications, a team of engineers supports locally the design-in and organizes the logistics by installed warehouses around the globe.

O. Bjoern
Sales & Marketing Manager
info@jianghai-europe.com
Adolf-Dembach-Str. 12
47829 Krefeld
Germany
Phone: 49 2151 652088-0
www.jianghai-europe.com

NIC COMPONENTS CORP.

Product: Aluminum Electrolytic Capacitors

NIC Components Corp. is a leading supplier of passive components, including an extensive line of SMT (V-Chip and Flat chip) and Leaded (Radial, Snap-In and Screw Terminal) aluminum electrolytic capacitors (E-caps). Our focus upon low ESR liquid, hybrid and solid electrolyte construction E-caps, allows us to offer the designer, more options and solutions than any other E-cap supplier. NIC has a direct presence in New York, California, Florida, Puerto Rico and Canada, including East and West Coast warehouses. Additional NIC European (NIC Components Europe) and South-East Asian (NIC Components Asia) divisions provide sales, product logistics and technical support to NIC customers in these regions. FREE "Design Your Own" Product Sample Kits at www.niccomp.com.

Eric Moller
National Sales Manager
sales@niccomp.com
70 Maxess Road
Melville, NY 11747
USA
Phone: (631) 396-7500
www.niccomp.com

KAIMEI ELECTRONIC CORP. (JAMICON)
13th. Fl. No. 81, Sec. 1, Hsin-Tai-Wu Road
Hsichih, 221 Taipei Hsien,
Taiwan, ROC
(88-622) 698-1010
(88-622) 698-0386 (Fax)

KENDEIL SRL
Via dell'Industria, 15/13
1-20020 Arese
(MI)
Italy
(39-02) 938-3378
(39-029) 358-0564 (Fax)

LECLANCHE SA
Capacitor Division
48 Avenue de Grandson
1400 Yverdon
Switzerland

LELON ELECTRONICS CORP.
20, Lane 51, Chen-Kung Rd.
Ta-Li City, Taichung Hsien
Taiwan
(88-642) 492-5858
(88-642) 492-2768 (Fax)

LUXON ELECTRONICS CORP.
No. 43-3, Beitau Village
Tamsui, Taipei Hsien
Taiwan
(886-42) 492-5858
(886-42) 492-2768 (Fax)

MAN YUE ELECTRONICS CO., LTD.
161 F, Yiko Industrial Building
10 Ka Yip Street
Chaiwan
Hong Kong
(85-22) 897-5277
Products: Radial Lead, Large Can, Axial Type VChip, Aluminum Electrolytic Capacitors.

THE PAUMANOK GROUP

Metal Oxide Varistors: World Market Outlook: 2008–2013

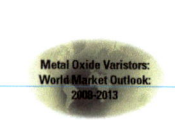

This market research report covers the global market for metal oxide varistors, including zinc oxide disc varistors, multilayered chip varistors (MLV), and chip varistor arrays. Global market demand is shown in terms of value and volume based on type (chip versus disc), application ESD versus overvoltage protection, demand by end-use market segment in wireless handsets, automotive electronic subassemblies, industrial power electronics, line voltage equipment and lighting ballasts, and other consumer electronics markets. Market growth is shown by year in value and volume from 1995–2008 by component configuration. Average selling prices and price erosion is also addressed over the past 13 years. The top 32 manufacturers of metal oxide varistors are discussed, and sales and market shares are given for 2008. Forecasts to 2013. 93 pages. Published December 2008.

Price: $1250 USD | **Additional Copies:** $125 USD
Downloadable (PDF) Copy: $1250 USD | **Additional License:** $125 USD
Product Code: MOV2008
Web: www.paumanokgroup.com/market_reports/ppf/c/4/reports.asp

For more details on these reports and others,
visit: www.paumanokgroup.com

SPRING ENGINEERING SUMMIT • APRIL 27-30, 2009 • NEW ORLEANS, LA

Our **2009 Spring Engineering Summit** is a four-day program designed to inform and connect leaders in the electronic components industry. This summit has a series of exciting and relevant general sessions and participation activities. It's your best chance to network with engineering leaders as they consider, develop and finalize crucial EIA Standards. EIA Standards set the precedence for safety, interoperability and performance optimization in an effective and efficient environment.

Don't miss your chance to:

- Contribute to your company's views on emerging standards
- Participate in streamlined standards development
- Promote new EIA Standards critical to your company's business
- Interface with industry experts on the issues that shape and define the market

> Make plans to be at the 2009 Spring Engineering Summit to ensure that your voice is heard on the technical course of the electronic components industry. Don't delay, register today!

Participation in Standards

Participation in Standards is open to all individuals who are directly and materially affected by matters addressed by committees, panels and working groups. Individuals from ECA member companies are eligible to participate in all EIA and ECA standards activities at no additional fee (except any special activities at ECA standards events). Individuals from non-ECA member companies and organizations are required to pay an annual per participant per committee fee of $1,000. This fee is in lieu of membership dues and supports administrative and overhead costs necessary to conduct ECA standards programs. Please see Attendance and Memberships information at www.ecaus.org/conference/spring_summit_2009/.

Awards

The Electronic Components Association (ECA) recognizes the hard work of its Engineering & Standards program participants. Recognize your peers and honor their achievements by submitting a nomination to ECA. The Electronic Component Engineering Awards will be presented at the Wednesday luncheon during the Spring Engineering Summit. The nominations are open to all members and participants in the ECA engineering activities.

Submit nominations by email to Cecelia Yates at cyates@ecaus.org or by faxing them to 703.875.8908. All nominations will be considered confidential unless otherwise designated by the submitter.

The awards descriptions are available on the web at **http://ecaus.org/conference/spring_summit_2009/** For more information or questions regarding the ECA Electronic Component Engineering Award Program, please call Cecelia Yates at 703.907.8026.

CAPACITORS

NICHICON (AMERICA) CORP.

Product: Aluminum Electrolytic Capacitors

Nichicon Corporation, a world renowned supplier of capacitors, provides market leadership and state of the art technology in Aluminum, Tantalum and Film capacitors. Specific state of the art Aluminum capacitors include our Pb-free, 260°C re-flow SMT products and SMT Polymer capacitors. Our tantalum capacitor line offers the state of the art, miniature size molded F98 series which offers cap. values up to 100 uf and up to 16 v in case codes "M" and "S." Our F95 conformal-coated tantalums are now extremely popular in circuits requiring high cap in limited space (especially in height restricted applications). Nichicon Film capacitors are available in both through-hole and also in our new "ML" SMT series. Nichicon is recognized as a global leader in products for the automotive industry including EDLC and Film capacitors for the Hybrid and Electric Vehicle markets.

All available products are RoHS compliant.

Mark Fisk
General Manager, Sales and Marketing
m.fisk@nichicon-us.com
927 E. State Parkway
Schaumburg, IL 60173
Phone: (847) 843-7500
www.nichicon.com

NIPPON CHEMI-CON CORP.
5-6-4 Osaki
Shinagawa-Ku, Tokyo
141-8605
Japan
(81-35) 436-7711
(81-35) 436-7631 (Fax)

PANASONIC ELECTRONIC DEVICES JAPAN CO.
1006 Kadoma
Kadoma City, Osaka 571-8506
Japan
(81-66) 908-1211
Products: MLCCs, MLC Leaded and SLCs.
Panasonic is primarily known as a blanket supplier of both fixed and variable on-board capacitors; they supply as much product by type, configuration and dielectric as they possibly can on a worldwide basis. Panasonic has been widely successful in DC film capacitors and aluminum electrolytic capacitors outside of Japan. They also produce ceramic capacitors and tantalum capacitors for captive consumption and merchant market sales.

RUBYCON AMERICA, INC.
4293 Lee Avenue
Gurnee, IL 60031
(847) 249-3450
Products: Aluminum Electrolytic and Film Capacitors

RUBYCON CORP.
Electrolytic Capacitor Division
1938-1, Nishi-Minowa Ina City
Nagano 399-4593
Japan
(026) 572-7111
(026) 573-2914 (Fax)
Products: Aluminum Electrolytic and Film Capacitors

SAMYOUNG ELECTRONICS CO., LTD.
146-1, Sangdaewon-Dong
Joongwon-Gu
Sungnam-City
Kyungki-Do, 462-807
Korea
(82-31) 740-2406
(82-31) 740-2427 (Fax)

SANYO ELECTRONIC DEVICE CO.
1-1 SANYO-cho, Daito, Osaka, 574-8534
Japan
(072) 872-0011
(072) 873-9858 (Fax)

SHOEI ELECTRONICS CO., LTD. (TAIYO YUDEN)
587-3 Sumiyoshi, Ueda-shi, Nagano 386-0002
Japan
(81-26) 822-3595
(81-26) 822-3559 (Fax)

TEAPO ELECTRONIC CORP.
No. 2, Lane 235, Alley 1
Baoqiao Rd., Xindian City, Taipei
Taiwan, ROC
(88-622) 910-5658
(88-622) 910-5352 (Fax)

TOBIAS JENSEN PRODUCTION A/S
Industrivej 4
2605 Brøndby
Denmark
(45-4) 327-1685
Products: Audio Signal Capacitors for Use in High-End Audio Amplifiers; Aluminum Foil Capacitors with Oil Impregnated Paper Dielectric; Metallized/Double Metallized Paper/Polyester/Polypropylene Capacitors with Mixed or Single Dielectric; High Voltage Capacitors with Mixed Paper-Polypropylene Dielectric Encased in Phenolic Paper Tube or Metal Housing; Middle and High Voltage Power Capacitors; Coupling Capacitors for Power Line Carrier (PLC) Systems.

TTI, INC.

Product: Never Short on Solutions

Headquartered in Fort Worth, Texas, TTI, Inc. is a distributor specialist of passive, interconnect, electromechanical, and discrete components. TTI is the distributor of choice for industrial and consumer electronic manufacturers worldwide.

TTI's product line includes: resistors, capacitors, connectors, potentiometers, trimmers, magnetic and circuit protection components, wire and cable, wire management, identification products, application tools, electromechanical devices, and discrete components. We distribute these products from a broad line of manufacturers.

TTI strives to be the industry's preferred information source by offering, through the ttiMarketEye blog, the latest IP&E technology and market information, technical seminars, RoHS seminars, industry research reports and much more.

TTI employs more than 2,000 people at more than 50 locations throughout North America, Europe, and Asia.

Sales Contact: information@ttiinc.com
2441 Northeast Pkwy
Fort Worth, Texas 76106
Phone: 800-CALL-TTI (225-5884)
www.ttiinc.com

CAPACITORS

UNITED CHEMI-CON, INC.

Product: Aluminum Electrolytic Capacitors

United Chemi-Con (a subsidiary of Nippon Chemi-Con) is recognized as one of the world's largest manufacturers of Aluminum Electrolytic Capacitors. Nippon Chemi-Con leads the capacitor industry in the production of aluminum electrode foil, the primary material that determines a capacitor's performance. With quality control over basic materials, as well as innovative research and manufacturing facilities (all ISO14001 certified), United Chemi-Con is capable of meeting customer demand for cost-effective, eco-friendly, high performance capacitors.

We continue to maintain a strong technological focus. We lead the way in polymer technology with our surface mount (PXA, PXE, PXF, PXH) and through-hole (PS, PSA, PSC, PSL) conductive polymer series capacitors. Our new multi-cell polymer capacitor Proadlizer® (WR Series) incorporates hundreds of "bubble" technology cells on one substrate for maximum capacitance per CV^3.

United Chemi-Con is promoting large can, snap-in, and radial tubular capacitors. Traditionally, our large can market has been strongly supported by our 36DA series capacitors, while our SMH and KMH series are still the stalwart of our snap-in products from our North Carolina facility. Our upgraded and newer featured series include:

- U36F, U36L, U32F series: Longer life
- U767D, U787D series: Higher vibration designs
- U82F, U81F series: Larger size, long life
- UTOR series: Higher CV per case size, reduced cost per amp, over-molded polymer mounting, post option

These upgraded series are a strong complement to our existing U32D and U36D large can series. Our engineers are ready to work with you to meet your large can design requirements.

As a company, we are dedicated to providing products and services that continually meet the needs and expectations of our customers. We want to be your source for Aluminum Electrolytic Capacitors!

Derrick Fitzpatrick
Manager Sales and Marketing
dfitzpatrick@chemi-con.com
9801 West Higgins Road
Rosemont, IL 60018
Phone: (847) 696-2000
Fax: (847) 696-9278
www.chemi-con.com

VISHAY INTERTECHNOLOGY, INC.
63 Lancaster Avenue
Malvern, PA 19355-2143
(610) 644-1300
(610) 296-0657 (Fax)
Products: Tantalum Capacitors (Molded Chip Tantalum Capacitors, Coated Chip Tantalum Capacitors, Leadframeless Chip Tantalum Capacitors, Through-Hole Solid Tantalum Capacitors, Wet Tantalum Capacitors, Capacitor Arrays and Modules); Ceramic Capacitors (Multilayer Chip Capacitors, Disc Capacitors); Film Capacitors; Power Capacitors; Heavy-Current Capacitors; Aluminum Capacitors; Silicon RF Capacitors.

Capacitor Brands:
Vishay BCcomponents
Vishay Cera-Mite
Vishay Draloric
Vishay ESTA
Vishay Roederstein
Vishay Sprague (includes Mallory [NACC] and Tansitor)
Vishay Vitramon
and others

THE PAUMANOK GROUP

PTC Thermistors: World Market Outlook: 2008–2013

This report addresses the global market for PTC thermistors in both ceramic and polymer construction, and in their various configurations—radial, surface mount, and axial strap, as well as their vertical integration into probe assemblies. The study shows a ten-year historical assessment of the global value and volume of consumption for PTC thermistors by type between 1998 and 2008 with forecasts to 2013. An analysis of price erosion for PTC thermistors by type is also given, with reasons for price erosion given by type. Markets for ceramic PTC thermistors are broken down into overcurrent protection, thermal overload protection, and temperature sensing; while markets for polymer PTC thermistors are broken out by overcurrent protection and thermal protection (batteries). A detailed end-market assessment is given for subscriber-line interface cards, DC motors, consumer electronics, and lighting ballasts. Demand by world region is discussed, with forecasts for each region to 2013. Sales and market shares for the top vendors by technology is given, as is a detailed look at each known vendor by technology (39 vendors overall). Technology assessment, trends, and directions are noted, and overall forecasts are given to 2013. 119 pages. Released December 2008.

Price: $1250 USD | **Additional Copies:** $125 USD
Downloadable (PDF) Copy: $1250 USD
Additional License: $125 USD | **Product Code:** PTC08
Web: www.paumanokgroup.com/market_reports/ppf/c/4/reports.asp

For more details on these reports and others, visit:
www.paumanokgroup.com

CAPACITORS

DC Film Capacitor Manufacturers

ARCOTRONICS GROUP (KEMET)
Via San Lorenzo, 19
40037 Sasso Marconi, Bologna
Italy
(39-05) 193-9111
(39-05) 184-0684 (Fax)
Products: DC Film Capacitors and AC Interference Suppression Capacitors.

AVX FRANCE
Av. Du Colonel Prat.
TPC 21850 St. Appollinaire
France
(33-8) 071-7400
Products: Major European Producer of DC Film Capacitors.

BISHOP ELECTRONICS CORP.
3729 B San Gabriel River Parkway
Pico Rivera, CA 90660-1483
(562) 695-0446
(562) 692-4008 (Fax)
Products: Metallized Polyester, Metallized Polycarbonate and Metallized Polypropylene; Customized Film Capacitor Manufacturing for Feedthrough and High Voltage Applications.

COMPONENT RESEARCH CO., INC.
1655 26th Street
Santa Monica, CA 90404
(310) 829-3615
Products: Specialty Supplier of DC Film Capacitors to the North American Market.
End-use markets include military/aerospace applications, but may be used in any commercial mission critical application, including medical and oil well logging (down-hole pump applications).

DEARBORN ELECTRONICS, INC.
1221 North Highway 17-92
Longwood, FL 32750
(407) 695-6562
Products: Film Capacitors for Switching Power Supplies in Computer and Computer Peripheral Applications and Hi-Rel Products for Military/Aerospace Applications in Aircraft Navigation, Instrumentation and Communications.

DEKI ELECTRONICS LIMITED
B-20, Sector 58
Noida 201 301 U.P.
India
(91-120) 258-4687
(91-120) 258-5289 (Fax)
Products: Plain Polyester, Metalized Polyester, Plain Polypropylene, Metalized Polypropylene, Plain & Metalized Polypropylene Mixed, Mixed Dielectric.

EFC/WESCO

Products: Film & Electrolytic Capacitors

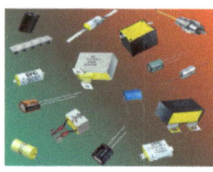

EFC/WESCO, your complete source for Film and Power capacitors, is proud to introduce their line of VDE and UL recognized EMI suppression capacitors and Capxon line of Electrolytics.

The EMI Suppression Films are "X" and "Y" class radial lead Polypropylene Box style with voltage ratings to 275 VAC in the "X2" class and capacitance to 3.3 µF.

Our Capxon line of Electrolytic capacitors offers all of the popular series including "Snap In" and "Surface Mount" packaging. Please contact the factory for our latest catalog.

EFC/WESCO is proud to be an ISO 9001:2000 certified company.

Bob Fountain
Sales Manager
fountain@filmcapacitors.com
41 Interstate Lane
Waterbury, CT 06705
Phone: 203-755-5629
www.filmcapacitors.com

ELECTROCUBE, INC.
3366 Pomona Blvd.
Pomona, CA 91768
(800) 515-1112
Products: Custom Film Capacitors.
Unique engineering and manufacturing capabilities in the design and creation of custom film capacitors. Acquired Seacor in 2002.

EPCOS AG
(EPCOS and TDK will merge in 2009)
P.O. Box 80 17 09
81617 Munich
Germany
(49) 896-3609
(49) 896-362-2689 (Fax)

EPCOS, INC.

Product: DC Film Capacitors and EMI Suppression Capacitors

EPCOS offers a complete line of DC Film Capacitors for general purpose, Pulse, Snubbering and RFI/EMI suppression Applications.
Metallized Polyester Film Capacitors cover a wide range of capacitance from 0.1-33µF. Metallized Polypropylene Capacitors, known for high voltage and pulse handling capabilities are rated up to 2KV in MKP (Metallized Film) construction and to 3KV in MFP (Metallized Film and Foil) construction. EPCOS also produces Capacitors in Tape wrapped Axial, Molded Box and Powder Dipped encapsulations.
In addition to their general purpose range, EPCOS offers DC Film Capacitors that are specially developed for use in Automotive, Industrial, Lighting and Power Supply applications. B3292x series is the latest X2 - EMI suppression capacitor series with UL Listing combines one of the smallest sizes in the industry along with high AC Voltage rating of 305 VAC and 125°C temperature.
EPCOS also offers special capacitors for Hybrid and Electric Vehicle drives.

Graciela Saldua
Product Marketing Engineer
graciela.saldua@epcos.com
186 Wood Avenue South
Iselin, NJ 08830
Phone: (800) 888-7729
www.epcos.com

EUROFARAD
93 rue Oberkampf
F-75540 Paris Cedex 11
France
(33-14) 923-1000
(33-14) 357-0533 (Fax)
Products: Metallized Film, Aluminum Electrolytic Capacitors, Tantalum and Ceramic Capacitors.

CAPACITORS

EVOX RIFA
A KEMET COMPANY

Product: Radial-Lead and SMD Film Capacitors, and Screw Terminal, Snap-In and Axial Electrolytics

Evox Rifa is a global supplier of radial-lead and SMD film capacitors, and screw terminal, snap-in and axial electrolytics. Many recent developments have been made in the area of DC Link capacitors for AC motor drives in industrial automation and in hybrid and electric vehicles. Evox Rifa can offer the optimum solution from a broad line of electrolytic and film capacitor options.

Evox Rifa's SMD film capacitors offer stable electrical properties and have been improved to offer even higher reflow soldering temperatures - up to 260ºC. Their encapsulated design offers superior environmental withstand. The electrodes are formed from flexible metal tabs to eliminate thermal stress cracking in even the largest sizes. Available in three dielectrics and up to 1000VDC.

One of the broadest lines of X and Y capacitors are also offered. One can find X capacitors in all industrial voltages up to 760VAC, eliminating the need to put two capacitors in series. Choose from economical plastic film or rugged impregnated paper. Both offer excellent self-healing properties. Evox Rifa also offers Y capacitors in plastic film, impregnated paper and ceramic.

All products are fully RoHS compliant and delivered from manufacturing facilities with ISO9000 and TS16949 certifications.

Evox Rifa, Inc.
capmaster@kemet.com
1640 Northwind Blvd., Unit 102
Libertyville, IL 60048
Phone: (847) 362-6770
www.kemet.com

FARATRONIC CO.
101 Jin Qiao Road
Xiamen 361012
China
(00-86-592) 533-1011
Products: Film Capacitors and Metallized Films.

F-DYNE ELECTRONICS CORP. (ELECTROCUBE, INC.)
Purchased by Electrocube, Inc. in 2007
1307 S. Myrtle Ave.
Monrovia, CA 91016
(800) 515-1112
(626) 357-8099 (Fax)

FUJITSU TOWA ELECTRON, LTD.
4-1-1 Kamikodanaka, Nakahara-ku
Kawasaki, Kanagawa 211-8588
Japan
(81-44) 777-1111

ITW PAKTRON

Products: Multilayer Polymer Film Capacitors

ITW Paktron manufactures Multilayer Polymer (MLP) Film Capacitors featuring ultra-low ESR and high ripple current capability. Designed for high frequency filtering and EMI/RFI suppression in AC/DC & DC/DC power conversion applications for telecommunications/datacom, military infrastructure, automotive, medical, and high-end industrial. The 100 V rated MLP capacitor is a standard choice for input/output filtering in 48 V telecom bus power applications (on-board or DC/DC modules). The MLP capacitor provides improved stability, both electrically and mechanically, compared to multilayer ceramics. The MLP features "non-shorting" operation and does not crack like large ceramic chip capacitors. Capacitance values range from 0.1 to 20 µF and voltage ratings 50 to 500 VDC, available in thru-hole or SMD mounting configurations. MLP trademarks include Capstick®, Angstor®, and Surfilm®. Paktron also offers the Quencharc® RC snubber networks, which are used to suppress inductive load electrical noise that can lead to contact arcing or resetting of PLCs.

Tom Saunders
Technical Sales Manager
itwpaktron@paktron.com
1205 McConville Road
Lynchburg, VA 24502
Phone: (434) 239-6941
www.paktron.com

OKAYA ELECTRIC INDUSTRIES CO., LTD.
6-16-9 Todoroki, Setagaya-ku, Tokyo 158-8543 Japan
(81-34) 544-7025
(81-34) 544-7055 (Fax)
Products: Specilizes in X&Y Suppression Film Capacitors.

PANASONIC ELECTRONIC DEVICES CO., LTD.
1006 Kadoma
Kadoma City, Osaka 571-8506
JAPAN
(81-66) 906-1652
Products: MLCCs, MLC Leaded and SLCs.
Panasonic is primarily known as a blanket supplier of both fixed and variable on-board capacitors; they supply as much product by type, configuration and dielectric as they possibly can on a worldwide basis. Panasonic has been widely successful in DC film capacitors and aluminum electrolytic capacitors outside of Japan. They also produce ceramic capacitors and tantalum capacitors for captive consumption and merchant market sales.

PILKOR ELECTRONICS CO., LTD.
381, Wonchun-Dong YeongTong-Ku
Suwon City, Gyunggi-Do
Korea
(82-31) 217-2500
(82-31) 217-7313 (Fax)
Products: One of the Largest Film Capacitor Producers in Korea.

RELIABLE CAPACITORS
12931 Sunnyside Place
Santa Fe Springs, CA 90670
(562) 946-8577
Products: Film and Foil Capacitors for High-End Audio.

RTI ELECTRONICS, INC.
1800 E. Via Burton St.
Anaheim, CA 92806-1213
(714) 765-8200
(714) 765-8201 (Fax)

RUBYCON CORP.
Electrolytic Capacitor Division
1938-1, Nishi-Minowa Ina City
Nagano 399-4593
Japan
(026) 572-7111
(026) 573-2914 (Fax)
Products: Electrolytic and DC Film Capacitors.

CAPACITORS

TTI, INC.

Product: Never Short on Solutions

Headquartered in Fort Worth, Texas, TTI, Inc. is a distributor specialist of passive, interconnect, electromechanical, and discrete components. TTI is the distributor of choice for industrial and consumer electronic manufacturers worldwide.

TTI's product line includes: resistors, capacitors, connectors, potentiometers, trimmers, magnetic and circuit protection components, wire and cable, wire management, identification products, application tools, electromechanical devices, and discrete components. We distribute these products from a broad line of manufacturers.

TTI strives to be the industry's preferred information source by offering, through the tti|MarketEye blog, the latest IP&E technology and market information, technical seminars, RoHS seminars, industry research reports and much more.

TTI employs more than 2,000 people at more than 50 locations throughout North America, Europe, and Asia.

Sales Contact: information@ttiinc.com
2441 Northeast Pkwy
Fort Worth, Texas 76106
Phone: 800-CALL-TTI (225-5884)
www.ttiinc.com

VISHAY INTERTECHNOLOGY, INC.
63 Lancaster Avenue
Malvern, PA 19355-2143
(610) 644-1300
(610) 296-0657 (Fax)
Products: Tantalum Capacitors (Molded Chip Tantalum Capacitors, Coated Chip Tantalum Capacitors, Lead-frameless Chip Tantalum Capacitors, Through-Hole Solid Tantalum Capacitors, Wet Tantalum Capacitors, Capacitor Arrays and Modules); Ceramic Capacitors (Multilayer Chip Capacitors, Disc Capacitors); Film Capacitors; Power Capacitors; Heavy-Current Capacitors; Aluminum Capacitors; Silicon RF Capacitors.

Capacitor Brands:
Vishay BCcomponents
Vishay Cera-Mite
Vishay Draloric
Vishay ESTA
Vishay Roederstein
Vishay Sprague (includes Mallory [NACC] and Tansitor)
Vishay Vitramon
and others

WIMA GMBH & CO.KG
WIMA Spezialvertrieb
elektronischer Bauelemente GmbH & Co.KG
Pfingstweidstr. 13
D-68199 Mannheim, Germany
(00-49-62) 186-2950
Products: Film/foil and metallized capacitors based on PET, PEN, PP, PPS, and mixed film as dielectrics.

CAPACITORS

AC Film Capacitor Suppliers

ABB LTD.
Affolternstrasse 44
CH-8050 Zurich
Switzerland
(41-043) 317-7111
(41-043) 317-4420 (Fax)

ADVANCE COMPONENTS & INSTRUMENTS PVT., LTD.
3A Belavadi Industrial Area
Mysore 570 018
India
(91-821) 240-2307
(91-821) 240-3058 (Fax)
Products: AC & DC Oil Filled Capacitors for Snubber & SCR Commutation. Also DC Film Leaded Products.

AEROVOX
Electrical Products Group
167 John Vertente Boulevard
New Bedford, MA 02745
(508) 994-9661
(508) 995-3000 (Fax)
Products: AC Film Capacitors.

AMBER CAPACITORS, LTD.
152 Ataturk Block,
New Garden Town, Lahore
Pakistan
(92-42) 584-3553
(92-42) 588-6635 (Fax)

ARCOTRONICS GROUP (KEMET)
Via San Lorenzo, 19
40037 Sasso Marconi, Bologna
Italy
(39-05) 193-9111
(39-05) 184-0684 (Fax)
Products: AC Interference Suppression Capacitors, AC-Oil Filled Capacitors and Winding Machines.

BYCAP, INC.
5505 N Wolcott Ave
Chicago, IL 60640
(773) 561-4976
Products: AC Oil-Filled Capacitors, DC Film Capacitors.

CAMEL TECHNOLOGY, INC.
No. 19 Youn Kong 2rd An Industrial Park
Kaohsiung
Taiwan
(88-67) 621-2131
(88-67) 622-4252 (Fax)
Products: AC and DC Film and Tantalum Capacitors.

**CHICAGO CONDENSER CORP.
(CAPACITOR INDUSTRIES)**
6455 N. Avondale Ave.,
Chicago, IL 60631
(773) 774-6666
(773) 774-6690 (Fax)
Products: High Voltage Oil Filled, Metallized Film, Capacitor Bypass Feedthrough, and Pulse Discharge Capacitors for Military and Commercial OEM Customers.

COMAR CONDENSATORI SPA
Via del Lavoro 80
40011 Anzola Emilia
Crespellano (Bologna)
Italy
(39-05) 173-3383
(39-05) 173-3620

CONDENSER PRODUCTS
2131 Broad St.
Brooksville, FL 34604
(352) 796-3561
(352) 799-0221 (Fax)
Products: High Voltage Oil Filled Capacitor Products for DC Filtering Energy Storage, Pulse Discharge Capacitors for Coupling, and Voltage Multipliers for OEM Applications.

COOPER POWER SYSTEMS
1319 Lincoln Avenue
Waukesha, WI 53186
(262) 524-3300
(262) 524-3319 (Fax)
Products: High Voltage Capacitors for power utilties in single-phase units; Pole Pounted Racks with single phase units or Block Banks at substations. These units are primarily 1kV and above.

CORNELL DUBILIER ELECTRONICS
140 Technology Place
Liberty, SC 29657
(864) 843-2277
(864) 843-3800 (Fax)

CSI CAPACITORS
2540 Fortune Way
Vista, CA 92081
(760) 682-2222
(760) 682-3333 (Fax)
Products: High-Voltage Paper and Paper Film Capacitors.

DEKI ELECTRONICS LIMITED
B-20, Sector 58
Noida 201 301 U.P.
India
(91-120) 258-4687
(91-120) 258-5289 (Fax)
Products: Plain Polyester, Metalized Polyester, Plain Polypropylene, Metalized Polypropylene, Plain & Metalized Polypropylene Mixed, Mixed Dielectric.

DUCATI ENERGIA SPA
Via Marco Emilio Lepido 182
40132 Bologna
Italy
(39-051) 641-1511
(39-05) 140-2040 (Fax)
Products: AC-Oil Filled Capacitors for Microwave Ovens and Power Factor Correction; Aluminum Electrolytic Capacitors for Motor Starting.

ELECTRONICON KONDENSATOREN GMBH
Keplerstrasse 2
07549 Gera
Thüringen, Germany
(49-365) 734-6100
(49-365) 734-6110 (Fax)
Products: AC Oil Filled Capacitors for Ballasts, Power Factor Correction, and Capacitor Banks.

EPCOS AG
(EPCOS and TDK will merge in 2009)
P.O. Box 80 17 09
81617 Munich
Germany
(49) 896-3609
(49) 896-362-2689 (Fax)

EVOX RIFA GROUP (KEMET)
Stella Business Park
Lars Sonckin kaari 16
02600 Espoo
Finland
(358-95) 406-5000
(358-95) 406-5010 (Fax)

FACON SPA
Via Molini Trotti
13 – 2100 VARESE
(39-33) 228-2300
(39-33) 228-2705 (Fax)
Products: AC-Oil Filled Capacitors, Polyester Capacitors and Aluminum Electrolytic Capacitors.

FARATRONIC CO.
101 Jin Qiao Road
Xiamen 361012
China
(00-86-592) 533-1011
Products: AC film capacitors for the application of lighting, motor start, and power systems.

FRAKO
Tscheulinstrasse 21a
D-79331 Teningen
Germany
(49-7) 641-4530
(49-764) 145-3535 (Fax)
Products: Metallized Paper/Film Capacitors (High Voltage).

GE CAPACITORS BY REGAL-BELOIT
1946 West Cook Road
Fort Wayne, Indiana 46818
(260) 416-5400
Products: High Voltage Power Film Capacitors for Power Transmission and Distribution.

GE COMMERCIAL MOTORS BY REGAL-BELOIT
1946 West Cook Road
Fort Wayne, IN 46818
(260) 416-5400
Products: Low Voltage Power Factor Correction Capacitors, Residential and Commercial Applications.

HIGH ENERGY CORP.
14 Lower Valley Road
Parkesburg, PA 19365
(610) 593-2800
(610) 592-2985 (Fax)
Products: Custom High Voltage Ceramic and Oil-Filled Capacitors for use in Military, Medical, and Industrial Capital Equipment.

HIVOLT CAPACITORS, LTD.
Maydown Industrial Estate, Derry
N. Ireland BT47 6UQ
(44-15) 486-0265
(44-15) 486-0479 (Fax)
Products: High Voltage AC & DC Oil Filled Capacitors, Plastic Leaded, Paper-Leaded Capacitors.

ICAR SPA
Via Isonzo 10
I-20052 Monza (MI)
Italy
(390) 398-3951
(39-03) 983-3227 (Fax)
Products: Paper/Film Capacitors.

CAPACITORS

JOHNSON & PHILLIPS CAPACITORS (MEM)
Eaton MEM
Reddings Lane
Birmingham B11 3EZ
United Kingdom
(01-21) 685-2100
(01-21) 685-2382 (Fax)
Products: AC Oil-Filled Capacitors to 11,000V.

KUMKANG ELECTRONICS CORP.
779-5, Chojon-ri
Chillye-Myon, Kimhae-kun
Kyongsanngnam-do
Korea
(82-52) 545-4101
(82 52) 545-4710 (Fax)
Products: DC Film Capacitors, RC Networks, RFI Interference Suppression Capacitors, and High Voltage Capacitors.

MFD CAPACITORS, LTD.
Lion Lane
Penley, Wrexham LL13 OLY
United Kingdom
(44-97) 871-0551
(44-0-197) 871-0501 (Fax)
Products: AC Oil-Filled and High Voltage Capacitors.

NUEVA GENERACION MANUFACTURAS, S.A. DE C.V.
Av Tezozomoc 239, Fracc. Ind.
San Antonio Azcapotzalco 02760
Mexico DF
(52-555) 352-5244
Products: Motor Start Capacitors: Phenolic Case, Max Start Case, and Mini Start Case. Motor Run Capacitors: Plastic Case, Plastic Box, Wet Metal Case Metallized, and Wrap & Fill. Axial and Radial Aluminum Electrolytic Capacitors. Hardware Accessories: Plastic Case, End Caps, and Brackets. Power Factor Correction: Capacitor Cells.

NWL INC.
8050 Monetary Drive
Riviera Beach, FL 33419 USA
(561) 848-9009
(561) 848-9011 (Fax)
Products: Film Capacitors.

NORFOLK CAPACITORS, LTD.
Leyden Works, Station Road
Great Yarmouth
Norfolk, NR31 0HB
United Kingdom
(44-49) 365-2752
(44-49) 365-5433 (Fax)
Products: AC and DC High Voltage and Pulse Discharge Capacitors.

RONKEN INDUSTRIES, INC.
9 Wolfer Industrial Park
Spring Valley, IL 61362
(815) 664-5306
Products: General purpose motor run marketplace.

SHIZUKI ELECTRIC CO., INC.
10-45 Taisha-cho
Nishinomiya, Hyogo 662-0867
Japan
(81-79) 874-5821
(81-79) 873-0807 (Fax)
Products: DC Film Capacitors, AC Interference Suppression Capacitors, and AC Oil-Filled Capacitors.

SPELLMAN HIGH VOLTAGE ELECTRONICS CORP.
475 Wireless Boulevard
Hauppauge, NY 11788
(631) 630-3000
Products: Power Conversion Systems.

TOBIAS JENSEN PRODUCTION A/S
Industrivej 4
2605 Brøndby
Denmark
(45-4) 327-1685
Products: Middle and High Voltage Power Capacitors; Coupling Capacitors for Power Line Carrier (PLC) Systems.

TTI, INC.

Product: Never Short on Solutions

Headquartered in Fort Worth, Texas, TTI, Inc. is a distributor specialist of passive, interconnect, electromechanical, and discrete components. TTI is the distributor of choice for industrial and consumer electronic manufacturers worldwide.

TTI's product line includes: resistors, capacitors, connectors, potentiometers, trimmers, magnetic and circuit protection components, wire and cable, wire management, identification products, application tools, electromechanical devices, and discrete components. We distribute these products from a broad line of manufacturers.

TTI strives to be the industry's preferred information source by offering, through the tti|MarketEye blog, the latest IP&E technology and market information, technical seminars, RoHS seminars, industry research reports and much more.

TTI employs more than 2,000 people at more than 50 locations throughout North America, Europe, and Asia.

Sales Contact: information@ttiinc.com
2441 Northeast Pkwy
Fort Worth, Texas 76106
Phone: 800-CALL-TTI (225-5884)
www.ttiinc.com

VISHAY INTERTECHNOLOGY, INC.
63 Lancaster Avenue
Malvern, PA 19355-2143
(610) 644-1300
(610) 296-0657 (Fax)
Products: Tantalum Capacitors (Molded Chip Tantalum Capacitors, Coated Chip Tantalum Capacitors, Leadframeless Chip Tantalum Capacitors, Through-Hole Solid Tantalum Capacitors, Wet Tantalum Capacitors, Capacitor Arrays and Modules); Ceramic Capacitors (Multilayer Chip Capacitors, Disc Capacitors); Film Capacitors; Power Capacitors; Heavy-Current Capacitors; Aluminum Capacitors; Silicon RF Capacitors.

Capacitor Brands:
Vishay BCcomponents
Vishay Cera-Mite
Vishay Draloric
Vishay ESTA
Vishay Roederstein
Vishay Sprague (includes Mallory [NACC] and Tansitor)
Vishay Vitramon
and others

WIMA GMBH & CO.KG
WIMA Spezialvertrieb
elektronischer Bauelemente GmbH & Co.KG
Pfingstweidstr. 13
D-68199 Mannheim, Germany
(00-49-62) 186-2950
Products: Self-healing, metallized capacitors for high pulse and high frequency applications.

WORLD PRODUCTS INC.
19654 Eighth Street East
Sonoma, CA 95476
(770) 996-5201
Products: RFI Capacitors

CAPACITORS

EDLC Supercapacitor Suppliers

ASAHI GLASS CO., LTD.
1-12-1 Yurakucho
Chiyoda-ku, Tokyo 100-8405
Japan
(81-33) 218-5741
(81-33) 218-7815 (Fax)

AVX CORP.
801 17th Avenue South
Myrtle Beach, SC 29578
(843) 448-9411
Product:: BestCap®

CAP-XX LTD.
9/12 Mars Road
Lane Cove NSW 2066
Australia
(61-29) 420-0690
(61-29) 420-0673 (Fax)
Products: Spiral Wound and Monoblock Supercapacitors that Employ a Carbon Particulate Material with a Binder and an Organic Electrolyte.

COOPER BUSSMANN ELECTRONICS
1225 Broken Sound Parkway NW
Boca Raton, FL 33487
(561) 998-4100
(561) 241-6640 (Fax)
Products: Commercialized Aerogel Carbon Technology for use in EDLC Capacitors.
This technology development can be traced to defense related research at Lawrence Livermore Labs.

ELECTROMECHANICAL RESEARCH, LTD. (AVX)
3 Hamarpe St. Har Hotzvim
P.O. Box 45188
Jerusalem 91450
Israel
(02) 572-0177

ELNA CO., LTD.
3-8-11 Shin-Yokohama
Kouhoku-ku
Yokohama-city 222-0033
Kanagawa
Japan
(81-45) 470-7254
(81-45) 470-7260 (Fax)
Products: Double Layer Carbon Supercapacitors for Consumer Electronic and Computer Applications in Japan, the United States, and Western Europe; also a Major Aluminum Electrolytic Producer.
Their parent company is Asahi Glass, Japan.

ESMA
JSC ESMA OKB FIAN
Troitsk
Moscow Region, 142190
Russia
Products: Nickel Hydroxide Supercapacitors.

EVANS CAPACITOR CO.
72 Boyd Avenue
East Providence
Rhode Island 02914
(401) 435-3555
(401) 435-3558 (Fax)
Products: Hybrid Supercapacitors that employ Conventional Anodes with Supercapacitor Cathode Materials to produce higher voltage per cell.
Evans hybrid designs are used in medical and military applications.

HITACHI MAXELL, LTD.
1-1-88, Ushitora
Ibaraki-shi, Osaka 567-8567
Japan
Products: Low Voltage Supercapacitors.

LECLANCHÉ CAPACITORS
1400 Yverdon
Switzerland
(41-24) 445-6688
(41-24) 445-6689 (Fax)
Parent Company: SCNE in France. LeClanche is jointly developing supercapacitors with ABB Corporate Research in Switzerland.

LS CABLE
555 Hogye-dong, Dongan-gu
Anyang-si, Gyeonggi-do 431-831
Korea
(82-31) 428-4719

MAXWELL TECHNOLOGIES
9244 Balboa Ave.
San Diego, CA 92123
(858) 503-3300
Products: Boostcap Ultracapacitors.

NEC CORP.
7-1, Shiba 5-chome
Minato-ku, Tokyo, 108-01
Japan
(83-33) 454-1111
Products: NEC Corporation has developed a High Voltage Supercapacitor. This product is developed separate from the business they sold to Tokin Corporation. The product is targeted toward the automotive industry in Japan.

NEC TOKIN, CORP.
Chiyoda First Bldg., 8-1, Nishi-Kanda 3-chome
Chiyoda-ku, Tokyo 101-8362
Japan
(81-33) 515-9222
(81-33) 515-9223 (Fax)
Products: A world powerhouse in tantalum capacitor manufacturing. Has developed supercapacitors based on DLC technology.

NESSCAP CO., LTD.
(446-901) 750-8 Gomae-dong, Giheung-gu
Yongin-si, Gyeonggi-do
Korea
(82-31) 289-0721
(82-31) 286-6767 (Fax)
Products: Metal Oxide Supercapacitor for applications in Battery Load Leveling of cellular phones.

NICHICON CORP.
Karasumadori Oike-agaru
Nakagyo-ku, Kyoto, 604-0845
Japan
(81-75) 231-8461
(81-75) 256-4158 (Fax)
Products: The company is primarily known for its Electrochemical Capacitors–Aluminum, Tantalum, and Niobium.
In 1999, the company announced a movement into the double layer carbon supercapacitor business, with bold production plans to produce some 30 million DLC capacitors in its first full year of sustained production.

NIPPON CHEMI-CON CORP.
5-6-4, Osaki
Shinagawa-Ku, Tokyo
141-8605
Japan
(81-35) 436-7711
(81-35) 436-7631 (Fax)

NISSAN MOTOR CO., LTD.
17-1 Ginza 6-chome
Chuo-ku, Tokyo 104-8023
Japan
(03-3) 543-5523

PANASONIC ELECTRONIC DEVICES CO., LTD.
1006 Kadoma, Kadoma City
Osaka 571-8506
Japan
(81-66) 906-1652
Products: MLCCs, MLC Leaded and SLCs.
Panasonic is primarily known as a blanket supplier of both fixed and variable on-board capacitors; they supply as much product by type, configuration and dielectric as they possibly can on a worldwide basis. Panasonic has been widely successful in DC film capacitors and aluminum electrolytic capacitors outside of Japan. They also produce ceramic capacitors and tantalum capacitors for captive consumption and merchant market sales.

SHIZUKI ELECTRIC
10-45 Taisha-cho, Nishinomiya-shi
Hyogo 622
Japan
(81-79) 874-5821
Products: Large Can Supercapacitor product line in Japan for use in automotive and related markets.

SIGMA TECHNOLOGIES INT'L, INC.
10960 N. Stallard Place
Tucson, AZ 85737
(520) 575-8013
Products: Hypercap ESDs.

CAPACITORS

TTI, INC.

Product: Never Short on Solutions

Headquartered in Fort Worth, Texas, TTI, Inc. is a distributor specialist of passive, interconnect, electromechanical, and discrete components. TTI is the distributor of choice for industrial and consumer electronic manufacturers worldwide.

TTI's product line includes: resistors, capacitors, connectors, potentiometers, trimmers, magnetic and circuit protection components, wire and cable, wire management, identification products, application tools, electromechanical devices, and discrete components. We distribute these products from a broad line of manufacturers.

TTI strives to be the industry's preferred information source by offering, through the tti|MarketEye blog, the latest IP&E technology and market information, technical seminars, RoHS seminars, industry research reports and much more.

TTI employs more than 2,000 people at more than 50 locations throughout North America, Europe, and Asia.

Sales Contact: information@ttiinc.com
2441 Northeast Pkwy
Fort Worth, Texas 76106
Phone: 800-CALL-TTI (225-5884)
www.ttiinc.com

UNITED CHEMI-CON, INC.

Product: DLCAP™ Electric Double Layer Capacitors

DLCAP™ Electric Double Layer Capacitors from United Chemi-Con represent the new frontier in capacitor technology and the current demand for renewable energy sources and "green" components. DLCAPs provide clean energy with activated carbon, use a non-acetonitrile electrolyte, are Pb-Free and Cd-Free, and are RoHS compliant. Combined in standard or custom modules or in racks of modules, large voltage options are possible for hybrid vehicles, aviation back-up energy, and high power assist and energy storage applications for wind turbines. Twenty standard models are currently available in a range of capacitance values. United Chemi-Con and our parent company Nippon Chemi-Con are dedicated to technology solutions that put the environment and people first.

We want to be your capacitor source!
- High Capacitance: 350 to 3,200 F (single cell)
- 2.3 V and 2.5 V Cylindrical and Prismatic Types
- Clean Energy with Activated Carbon
- Pb-Free, Cd-Free, Non-Acetonitrile Electrolyte
- Low Internal Resistance: 0.7 mΩ typ. (2,400 F)
- Over 1 Million Charge/Discharge Cycles
- Low Temperature Charge/Discharge (-30°C)
- Five Standard Modules Available (13.8 V and 15 V)
- Two Standard Racks of Modules Available (105 V and 210 V)

Derrick Fitzpatrick
Manager Sales and Marketing
dfitzpatrick@chemi-con.com
9801 West Higgins Road
Rosemont, IL 60018
Phone: (847) 696-2000
Fax: (847) 696-9278
www.chemi-con.com

WIMA GMBH & CO.KG / GERMANY

Product: WIMA SuperCap Capacitors

The Double-Layer capacitors developed by WIMA / Germany are storage capacitors with highest capacitance values in the Farad range. They are among others suited to serve as batteries, can deliver extremely high currents for a short time, and are maintenance-free. WIMA Double-Layer capacitors are available in the standard capacitance range of 100 F to 400 F with a rated voltage of 2.5 VDC and the highest discharge current. The prismatic aluminum case makes space-saving series and parallel connections possible. Unhelpful cavities are avoided. WIMA SuperCaps replace, protect, or support batteries in the field of new traction technologies e.g. in automotive, railway systems, wind power mills or in uninterruptible power supplies (UPS).
WIMA has been certified in accordance with ISO 9001 and meets the RoHS requirements.

Andreas (Andy) Kunz
Key Account Manager Export
andreas.kunz@wima.de
WIMA Spezialvertrieb
elektronischer Bauelemente GmbH & Co.KG
Pfingstweidstr. 13
D-68199 Mannheim
Germany
Phone: 0049-621-86295-0
www.wima.com

CAPACITORS

Niobium Capacitor Suppliers

AVX CORP.

Product: Niobium Oxide Capacitors (OxiCap®)

AVX developed a family of niobium oxide* capacitors designated the OxiCap® range. These capacitors utilize advanced technology to provide design engineers with a safe (non-burning) high resistance failure mode. A family of multi-anode, ultra low ESR devices (NOM series) offer high capacitance and high ripple current and capacitance values up to 680uF.
Ideal for power supply decoupling and high speed data processing applications, these capacitors provide excellent electrical characteristics with ESR levels, down to 23 milliohms.
OxiCap® offers reduced voltage derating and is an ideal choice to fill the performance gap between ceramic multilayer capacitors and aluminum electrolytics.

*Niobium Oxide capacitors are made under license from Cabot Corporation.

Dan Lane
Marketing Manager
dlane@avxus.com
(843) 946-0483
www.avxus.com

TTI, INC.

Product: Never Short on Solutions

Headquartered in Fort Worth, Texas, TTI, Inc. is a distributor specialist of passive, interconnect, electromechanical, and discrete components. TTI is the distributor of choice for industrial and consumer electronic manufacturers worldwide.

TTI's product line includes: resistors, capacitors, connectors, potentiometers, trimmers, magnetic and circuit protection components, wire and cable, wire management, identification products, application tools, electromechanical devices, and discrete components. We distribute these products from a broad line of manufacturers.

TTI strives to be the industry's preferred information source by offering, through the tti|MarketEye blog, the latest IP&E technology and market information, technical seminars, RoHS seminars, industry research reports and much more.

TTI employs more than 2,000 people at more than 50 locations throughout North America, Europe, and Asia.

Sales Contact: information@ttiinc.com
2441 Northeast Pkwy
Fort Worth, Texas 76106
Phone: 800-CALL-TTI (225-5884)
www.ttiinc.com

VISHAY INTERTECHNOLOGY, INC.
63 Lancaster Avenue
Malvern, PA 19355-2143
(610) 644-1300
(610) 296-0657 (Fax)
Products: Tantalum Capacitors (Molded Chip Tantalum Capacitors, Coated Chip Tantalum Capacitors, Leadframeless Chip Tantalum Capacitors, Through-Hole Solid Tantalum Capacitors, Wet Tantalum Capacitors, Capacitor Arrays and Modules); Ceramic Capacitors (Multilayer Chip Capacitors, Disc Capacitors); Film Capacitors; Power Capacitors; Heavy-Current Capacitors; Aluminum Capacitors; Silicon RF Capacitors.

Capacitor Brands:
Vishay BCcomponents
Vishay Cera-Mite
Vishay Draloric
Vishay ESTA
Vishay Roederstein
Vishay Sprague (includes Mallory [NACC] and Tansitor)
Vishay Vitramon
and others

NEC TOKIN CORP.
Chiyoda First Bldg., 8-1, Nishi-Kanda 3-chome
Chiyoda-ku, Tokyo 101-8362
Japan
(81-33) 515-9222
(81-33) 515-9223 (Fax)
Products: NEC Tokin Corporation has Developed Niobium Capacitor Technology In-House.
The company has worked very closely on employing polymer technology in the niobium capacitor cathode to lower ESR and make its performance closer to that of tantalum.

RESISTORS

Chip Resistor & Array

ABCO ELECTRONICS CO., LTD.
5448-4, Sangdaewon-dong
Jungwon-gu
Seongnam-city, Kyeonggi-Province
Korea
(82-31) 730-5000
(82-31) 743-2824 (Fax)
Products: Chip-Resistor, Metal Film.

BEIHAI YINHE HI-TECH IND. CO., LTD.
Yinhe Technology Building
Guangdong Road
Beihai
Guangxi 536000
China
(86-0779) 320-2636
(86-0779) 320-1888 (Fax)
Products: One of the largest indigenous Chip Resistor manufacturers in China.

CHAOZHOU THREE-CIRCLE CO., LTD.
Sanhuan Industrial District
Fengtang Chaozhou City 515646
Guangdong China
(86-0768) 685-9262
(86-0768) 685-5921 (Fax)
Products: Primarily a manufacturer of Alumina Rods for Axial and Radial Leaded Chip Resistor Production.

COMPOSTAR TECHNOLOGY
No. 7, Ta Yeh Street
Ta Fa Industrial District
Da Liao Hsiang
Kaohsiung 831
Taiwan
(88-67) 787-0611
(88-67) 788-0496 (Fax)
Products: Chip Resistors and Multichip Arrays.

COSONIC ENTERPRISE CORP.
6F-4, No. 77
Keelung Road, Sec. 2
Taipei
Taiwan
(88-62) 738-2233
(88-62) 738-5588 (Fax)
Products: Axial and Radial Leaded Carbon Film, Metal Film Resistors and a small line of Thick-Film Chip Resistors.

DONGGUAN SHIHHAO ELECTRONICS CO.
Chigang Industrial Park
Humen Town
Dongguan
Guangdong 523009
China
Products: Specialty Resistors; including Cement, Wirewound, Carbon Film, and Composition and Precision. Also has a line of Thick-Film Chip Resistors.

EPCOS AG
(EPCOS and TDK will merge in 2009)
P.O. Box 80 17 09
81617 Munich
Germany
(49) 896-3609
(49) 896-362-2689 (Fax)
Products: SMD PTC thermistors for power line voltages and CeraDiode® arrays for ESD protection of two USB 2.0 ports.

EPCOS, INC.
(EPCOS and TDK will merge in 2009)
186 Wood Ave. South
Iselin, NJ 08830
(800) 689-3717
Product: CeraDiodes.

EVER OHMS TECHNOLOGY CO., LTD.
No. 143, Lane 323 Ta-Liao Road
Ta-Liao Hsiang
Kaohsiung Hsien
Taiwan
(88-67) 788-4328
(88-67) 788-4320 (Fax)
Products: Thick-Film Chip Resistor producer in Taiwan.

FENGHUA ADVANCED TECHNOLOGY (GROUP) CO.
18th Fenghua Road
Zhaoqing City, Guangdong Province
China
(86-758) 286-5325
(86-758) 286-5174 (Fax)
Products: MLCC and Thick-Film Chip Resistors.

HAN RYUK ELECTRONICS CO., LTD.
785-4 Wonsi-Dong, Ansan-City
Kyungki-Do
Korea 425-090
(82-31) 493-8811
(82-31) 493-8805 (Fax)
Products:: Thick-Film Chip Resistor.

HMR CO., LTD.
159-43 Sosa bon dong, Sosa-Gu
Buchon-city, Kyunggi-Do 422-811
Korea
(82-32) 346-2211
(82-32) 348-7200 (Fax)
Products: Wirewound; Some Chip Resistors. Joint venture with Vishay-Dale.

HOKURIKU ELECTRIC INDUSTRY CO., LTD.
3158 Shimo-okubo, Toyama City
Toyama Pref. 939-2292
Japan
(076) 467-1111
(076) 468-1508 (Fax)
Products: High Voltage Thick-Film Chip Resistors.

IRC

Products: Resistive and Thermal Management

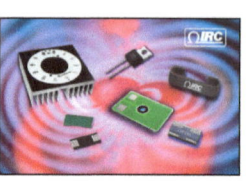

IRC offers an extensive range of resistor solutions including current sense resistors, precision discretes and networks, integrated passive components, and specialized power resistors. Resistive technologies include thin film tantalum nitride discretes and networks on ceramic and silicon substrates for precision high-frequency applications, thick film low- and high-resistance chips, wirewound resistors and assemblies, and cylindrical surface mount resistors. The company also offers a variety of thermal management products, including instantaneous heating solutions on Thick Film on Steel technology, as well as cooling solutions employing anodized aluminum substrates.

Charlotte Waters
Sales and Marketing Manager
afdsales@irctt.com
4222 S. Staples St
Corpus Christi, TX 78411
Phone: (361) 992-7900
www.irctt.com

ISABELLENHÜTTE HEUSLER GMBH & CO. KG
Eibacher Weg 3-5
D-35664 Dillenburg
Germany
(49-0-2) 771-9340
Products: Surface-mount passive components such as precision and power resistors.

KAMAYA ELECTRIC CO., LTD. (WALSIN)
A-209 KSP R&D Business Park Building
3-2-1 Sakado, Takatsu-ku, Kawasaki-shi, Kanagawa
213-0012
Japan
(81-44) 820-8560
(81-44) 820-8563 (Fax)
Products: One of the top producers of Thick-Film Chip Resistors in the world. This company is considered innovative in producing Ultra-Small Case Size Chip Resistors.

KOA CORP.
2-17-2 Midori-Cho
Fuchu-Shi
Tokyo 183-0006
Japan
(81-42) 336-5755
(81-42) 336-5353 (Fax)
Products: Thick-Film Chip Resistors in its Aichi, Tohoku, Takumino-Sato, Shangai, and Malaysian Production Facilities.

Continued on page 32

THE PAUMANOK GROUP

Get the latest in passive component market research from Paumanok Publications

Tantalum Global Market Outlook: 2008–2013

This study covers tantalum ore, engineered tantalum materials, tantalum capacitors, superalloys, cemented carbides, coatings and liners, lighting and optics, and emerging makets for tantalum.

The study forecasts demand for tantalum based upon its unique attributes of capacitance, durability, corrosion and heat resistance, as well as its unique optical properties to 2013. Key findings include issues facing the industry between 2008 and 2013, including the imminent price increases for the metal, its potential deficit supply, the forecast data for demand in capacitor anodes, cemented carbide cutting tools, nickel-based superalloys, wear parts, coatings and liners, and its two unique markets in tantalum wire and tantalum targets. You will also find:
- Forecasted price increases for tantalum to 2013.
- Forecasted deficits in ore supply to 2013.
- Shifts in the global market for ores and concentrates.
- Powder inventories in the supply chain.
- Volume of consumption requirements for capacitors, superalloys, cemented carbides, and other key end markets to 2013.
- Trends and directions in capacitors, superalloys, cutting tools, and wear parts.
- Complete lists of players at each profit center in the tantalum supply chain.
- Outlook for electronics, aircraft, automobiles, oil and gas, mining, power generation, and impact on tantalum demand to 2013.

Price: $2500 USD | **Additional Copies:** $250 USD
Downloadable (PDF) Copy: $2500 USD | **Additional License:** $250 USD
Product Code: TA20508
Web: www.paumanokgroup.com/market_reports/ppf/c/5/reports.asp

Capacitor Foil Global Market Outlook: 2008–2013

A global analysis of thin foil stock and etched anode and cathode foils consumed in the global aluminum electrolytic capacitor industry, with forecasts to 2013.

This study addresses the global supply chain for electrolytic foils consumed in capacitors. This is a narrow segment of the overall aluminum market, but is considered a value-added and application-specific portion of the supply chain for aluminum metal.
The study forecasts demand for thin dielectric aluminum foils used in capacitor anodes and cathodes to 2013 based upon consumption in various types and configurations of aluminum capacitors. The study also addresses the competitive nature of aluminum as a dielectric when compared with competing technologies.
The study addresses key issues that will face the industry between 2008 and 2013, including the impact of raw material price increases on capacitor manufacturers, and how competition for aluminum from multiple industries and regions will continue to impact the cost of goods sold in the capacitor industry over time. You will also find:
- Forecasted price increases for thin aluminum foil to 2013.
- Forecasted value and volume increases for etched anode and cathode foils to 2013.
- Trends in thin foil production.
- Trends in captive etching of foil.
- Trends in capacitor production.
- Competitive environment in foil supply to the capacitor industry.
- Market forecasts for aluminum foil consumption in the capacitor industry to 2008.

Price: $2500 USD | **Additional Copies:** $250 USD
Downloadable (PDF) Copy: $2500 USD | Additional License: $250 USD
Product Code: AL203FOIL
Web: www.paumanokgroup.com/market_reports/ppf/c/5/reports.asp

For more details on these reports and others, visit:
www.paumanokgroup.com

RESISTORS

Continued from page 30

KOA SPEER ELECTRONICS
199 Bolivar Drive
Bradford, PA 16701
(814) 362-5536
(814) 362-8883 (Fax)
Products: Current sensing resistors.

MICRO-OHM CORPORATION
1088 Hamilton Road
Duarte, CA 91010
(626) 357-5377
(626) 358-6478 (Fax)
Products: Resistors. New metal plate current sensing shunt resistor, type BHR and HPR, for automotive, motor control, and instrumentation.

MINI-SYSTEMS, INC.
20 David Road
PO Box 69
North Attleboro, MA 02761
(508) 695-0203
(508) 695-6076 (Fax)
Products: Thick/thin film chip resistors, networks, arrays, terminations, MOS capacitors, attenuators.

MMC ELECTRONICS AMERICA (KAMAYA)
1314-B N. Plum Grove Road
Schaumburg, IL 60173
(847) 252-6360
(847) 519-1736 (Fax)
Products: Thick film technology resistors. Also produces resistors with carbon composition, fusing, and high voltage.

NIC COMPONENTS CORP.

Product: Thick Film Resistors

NIC Components Corp. is a leading supplier of passive components, including a comprehensive line of precision SMT Thick Film and Thin Film Chip Resistor and Array products. Resistance values range from 0.05 ohm to 10 Meg-ohm in tolerances from ±0.1% to ±5%. Case sizes from 0201 to 2512 cover power ratings from 0.05W to 3.0W and temperature coefficients of ±5PPM to ±350PPM. 2-, 4- and 8-element thick film chip resistor arrays and zero-ohm jumper styles are also available. NIC continues to support leaded style carbon film, metal film and metal-oxide film resistors. NIC has a direct presence in New York, California, Florida, Puerto Rico and Canada, including East and West Coast warehouses. Additional NIC European (NIC Components Europe) and South-East Asian (NIC Components Asia) divisions provide sales, product logistics and technical support to NIC customers in these regions. FREE "Design Your Own" Product Sample Kits at www.niccomp.com.

Eric Moller
National Sales Manager
sales@niccomp.com
70 Maxess Road
Melville, NY 11747 USA
Phone: (631) 396-7500
www.niccomp.com

PANASONIC ELECTRONIC DEVICES CO., LTD.
1006 Kadoma
Kadoma City
Osaka 571-8506
Japan
(81-66) 906-1652
Products: MLCCs, MLC Leaded and SLCs.
Panasonic is primarily known as a blanket supplier of both fixed and variable on-board capacitors. Panasonic has been widely successful in DC film capacitors and aluminum electrolytic capacitors outside of Japan. They also produce ceramic capacitors and tantalum capacitors for captive consumption and merchant market sales.

PROSPERITY DIELECTRIC CO. (WALSIN)
10F, No. 480, Rueiguang Rd.
Neihu Chiu 114, Taipei
Taiwan
(886-22) 797-6698
(886-22) 797-1766 (Fax)
Products: MLCC, Chip-R, Ceramic Dielectric Powders, Disc-Type Semi-Conductive Capacitor Elements.

RALEC ELECTRONICS CORP.
8F-13, No. 79, Sec.1, Hsin Tai Wu Rd.
Hsi-Chih, Taipei Hsien
Taiwan
(88-62) 698-9977
(88-62) 698-9900 (Fax)
Products: Large producer of Chip Resistors in Taiwan.

ROHM CO., LTD.
21, Saiin Mizosaki-cho, Ukyo-ku
Kyoto 615-8585
Japan
(81-75) 311-2121
(81-75) 315-0172 (Fax)
Products: ROHM is one of the largest producers of Thick-Film Chip Resistors in the world.

SICHUAN YONGXING ELECTRONICS CO., LTD.
98 Electron Road, Xindu
Chengdu, Sichuan
China 610500
(86-28) 396-2201
(86-28) 396-5344 (Fax)

SRT RESISTOR TECHNOLOGY GMBH

Product: Resistors for Special Applications

SRT Resistor Technology is producing several series of chip resistors for special applications in its facility in Germany. The main product group includes high value chip resistors from sizes 0402 to 4020 with values up to 10 Teraohm, close tolerances, low values of TCR and VCR, and working voltages up to 6000 V. For higher voltages up to 30 kV, there are leaded types with high stability and TCR25 up to 1 Teraohm. Another important product group contains chip resistors with non-magnetic contacts (PtAg), which are required for medical applications using high magnetic fields like in CT or NMR devices. These types can also be used for conductive gluing for applications using wire-bonding. A similar version is suited for usage in high temperature applications up to +300°C. Other series of chip resistors can be used for functional trimming or as temperature sensors with PTC- or NTC-characteristics. Further types of resistors are supplied from international cooperation partners.

Dr. Lutz Baumann
Managing Director
Ostlandstr. 31
D 90556 Cadolzburg
Germany
info@srt-restech.de
Phone: ++49 9103 79520
www.srt-restech.de

RESISTORS

STACKPOLE ELECTRONICS, INC.
2700 Wycliff Road
Raleigh, North Carolina 27607
(919) 850-9500
(919) 850-9504 (Fax)

STATE OF THE ART, INC. (SOTA)
2470 Fox Hill Road
State College, PA 16803-1797
(800) 458-3401
(814) 355-8004
(814) 355-2714 (Fax)
Products: Thick and Thin-Film Chip Resistors and Networks (Mil-Spec).

SYNTON-TECH CORP.
16F-3, No 79, Far East World Center
Sec. 1, Hsin Tai Wu Rd, Hsi-Chih
Taipei County, Taiwan
(88-62) 698-1011
(88-62) 698-1012 (Fax)
Products: Cement Resistors, Carbon Film, and Carbon Composition, Tin Oxide, and Nichrome Film

TAD ELECTRONICS
9 Godstow Rd, Wolvercote
Oxford, OX2 8AJ, United Kingdom
(44-0-771) 724-3268
(44-0-870) 140-0160 (Fax)

TAI ELECTRONICS CO., LTD.
No.133, Nan-Kann Road
Sec 1 Lu-Jwu, Taoyuan, Taiwan
(886-03) 322-1988
(866-03) 352-4879 (Fax)
Products: Emerging as a major producer of Thick-Film Chip Resistors in Taiwan.

TTI, INC.

Product: Never Short on Solutions

Headquartered in Fort Worth, Texas, TTI, Inc. is a distributor specialist of passive, interconnect, electromechanical, and discrete components. TTI is the distributor of choice for industrial and consumer electronic manufacturers worldwide.

TTI's product line includes: resistors, capacitors, connectors, potentiometers, trimmers, magnetic and circuit protection components, wire and cable, wire management, identification products, application tools, electromechanical devices, and discrete components. We distribute these products from a broad line of manufacturers.

TTI strives to be the industry's preferred information source by offering, through the tti|MarketEye blog, the latest IP&E technology and market information, technical seminars, RoHS seminars, industry research reports and much more.

TTI employs more than 2,000 people at more than 50 locations throughout North America, Europe, and Asia.

Sales Contact: information@ttiinc.com
2441 Northeast Pkwy
Fort Worth, Texas 76106
Phone: 800-CALL-TTI (225-5884)
www.ttiinc.com

TYCO ELECTRONICS CORPORATION
1050 Westlakes Drive
Berwyn, PA 19312
(610) 893-9800

VISHAY INTERTECHNOLOGY, INC.
63 Lancaster Avenue
Malvern, PA 19355-2143
(610) 644-1300
(610) 296-0657 (Fax)
Products: Foil Resistors; Film Resistors (Metal Film Resistors, Thin Film Resistors, Thick Film Resistors, Metal Oxide Film Resistors, Carbon Film Resistors); Wirewound Resistors; Power Metal Strip® Resistors; Chip Fuses; Variable Resistors (Cermet Variable Resistors, Wirewound Variable Resistors, Conductive Plastic Variable Resistors); Networks/Arrays; Non-Linear Resistors (NTC Thermistors, PTC Thermistors, Varistors).
Resistor Brands:
Vishay Angstrohm
Vishay Beyschlag
Vishay BCcomponents
Vishay Dale
Vishay Draloric
Vishay Electro-Films
Vishay Foil Resistors
Vishay Phoenix
Vishay Sfernice
Vishay Spectrol
Vishay Techno
Vishay Thin Film
and otherrs

WALSIN TECHNOLOGY CORP.
566-1, Kao-Shi Rd
Yang-Mei, Tao-Yuan
Taiwan
(88-63) 475-8711
(88-63) 475-7129 (Fax)
Products: Chip resistors and arrays (SMD type).

YAGEO CORP.
3F No. 233-1 Pao-Chiao Rd
Hsin Tien, Taipei
Taiwan
(886-22) 917-7555
(886-22) 917-3789 (Fax)
Products: Yageo is the world's largest producer of Thick-Film Resistor Chips, with an Annual Capacity of 200 billion units.
Factories are located in Taiwan and China.

RESISTORS

Resistor Network Suppliers SIPs, DIPs, & IPD

ABCO ELECTRONICS CO., LTD.
5448-4, Sangdaewon-dong
Jungwon-gu
Seongnam-city, Kyeonggi-Province
Korea
(82-31) 730-5000
(82-31) 743-2824 (Fax)
Products: Resistor Networks.

BARRY INDUSTRIES, INC.
60 Walton St.
Attleboro, MA 02703
(508) 226-3350
(508) 226-3317 (Fax)
Products: Specialty producer of Thick and Thin-Film Resistor and R/C Networks. Also produce Microwave Resistors for defense and other electronic applications, and substrates (Including Teflon) for the Resistor Industry.

BI TECHNOLOGIES (TT ELECTRONICS)

Products: Thick and Thin Film Networks

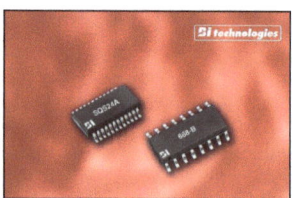

BI Technologies (TT electronics) produces both thick and thin film surface mount and through hole products. We supply products from +/-5% precision to +/-0.05% ultra precision networks. Temp coefficients range from +/-100 ppm to +/-25 ppm. Our products are all robust and capable of withstanding difficult environments and severe manufacturing processes.
To order or for further information, contact our customer service department at 714-447-2345 or visit our website.

sales@bitechnologies.com
4200 Bonita Pl
Fullerton CA 92835
Phone: (714) 447-2345
www.bitechnologies.com

BOURNS, INC.
1200 Columbia Avenue
Riverside, CA 92507
(951) 781-5690
(951) 781-5273 (Fax)
Products: A private company, specializing in production of Variable Resistors and Potentiometers. Also a large producer of SIP and DIP Thick-Film Networks.

CADDOCK ELECTRONICS, INC.

Products: Precision Resistor Networks

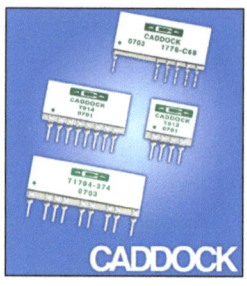

Caddock Electronics manufactures high performance film resistor networks in SIP packages for the most demanding precision analog applications.
- Transient Tolerant Precision Resistor SIP Network products -- for Power Quality Monitoring, Power Metering, and other applications with harsh electrical environments that demand precision.
- Resistor Pairs and Quads with Ultra-Precision Ratio performance in SIP packages. Ratio Tolerance to 0.01% and Ratio TC to 2ppm/°C.
- Decade Voltage Dividers – Input resistance of 10 Meg or 1 Meg, ratios of 10,000:1 to 10:1, input voltage rating to 1200 volts, precision ratio performance.
- High Voltage Dividers – Type HVD Resistor Networks with voltage ratings up to 5KV, Ratio Tolerance of 0.05% and Ratio TC 5ppm/°C.
- Custom SIP Resistor Networks are available in resistance values from 0.10 Ohms to 50 Megohms. Absolute tolerances from 0.025% to 1%, Ratio tolerances from 0.01% to 1%. Absolute TC's from 15ppm/°C to 50ppm/°C, Ratio TC's from 2ppm/°C to 50ppm/°C.

Applications Engineering
caddock@caddock.com
17271 N Umpqua Hwy
Roseburg OR 97470
Phone: (541) 496-0700
www.caddock.com

CALIFORNIA MICRO DEVICES
490 N. McCarthy Blvd. #100
Milpitas, CA 95035-5112
(408) 263-3214
(408) 263-7846 (Fax)
Products: Integrated Passive Devices based on Thin-Film Technology.

CINETECH RESISTOR CO., LTD.
7th Floor, No.10, Lane 235
Bao Chiao Road, Hsin Tien
Taipei Hsien
Taiwan
(88-20) 915-9601
Products: Thin-Film Integrated Passive Devices.

C-MAC MICROTECHNOLOGY
Endeavour House
Mercury Park, Wycombe Lane
Wooburn Green, Buckinghamshire
HP10 0HH
United Kingdom
(44-0-162) 885-9680
(44-0-162) 885-9690
Products: Major producer of Line Feed Resistor Networks for secondary Line Card protection in Telecommunications SLIC.

CTS CORP.
905 West Boulevard North
Elkhart, IN 46514
(574) 293-7511
(574) 293-6146 (Fax)
Products: Major producer of Thick-Film SIP and DIP Resistor Networks. Also a major producer of LTCC Components and modules.

DANAM CORP.
Mega Valley #701-2, 799 KwanYang-Dong,
DongAn-Ku
Anyang, KyungGi-Do 431-804
Korea
(82-31) 420-5951
(82-31) 420-5950 (Fax)
Products: Carbon Film and some Resistor Networks.

DIABLO INDUSTRIES, INC.
2245 Meridian Boulevard, Suite E
Minden, NV 89423
(702) 782-1041
Products: Thin-Film Resistors and Dual Plate Capacitors.

DSC ELECTRONICS CO., LTD
9-1 Suyoung-li, Bongdam-eup
Hwasung-Kun, Kyonggi-do
Korea
(82-31) 296-7093
(82-31) 296-7096 (Fax)

EBG ELEKTRONISCHE BAUELEMENTE
A-8082 Kirchbach 384
Austria
(43-31) 162-6250
(43-3) 116-2076 (Fax)
Products: Metal Film, Metal Oxide, Flat Chips, and Resistor Networks.

FIRST RESISTOR & CONDENSER CO., LTD.
9F. No.233, Sec. 4, Sinyi Rd.
Da-an District, Taipei City 106
Taiwan
(02) 705-1878
Products: Carbon Film, Metal Film, Metal Oxide, Flat Chip, Wirewound, and Resistor Networks.

FIRSTRONICS COMPONENTS
1020 Tai Seng Avenue
#04-3508 Tai Seng
Industrial Estate
Singapore 534416
(656) 287-3825
(656) 284-0368 (Fax)
Products: Carbon Film, Metal Film, and Resistor Networks.

RESISTORS

FUKUSHIMA FUTABA ELECTRIC CO., LTD.
12-14, Ikegami 8-Chome
Ota-Ku, Tokyo, 146-0082
Japan
(03-5) 700-3611
(03-5) 700-3343 (Fax)
Products: Resistor Networks, Wirewound, and Metal Film.

HOKURIKU ELECTRIC IND. CO., LTD.
3158 Shimo-okubo, Toyama City
Toyama Pref. 939-2292
Japan
(81-76) 467-1111
Products: Resistor Networks, Flat Resistor Chips, Variable Resistors, and Thermistors.

INTERNATIONAL MFG SERVICES (IMS)
50 Schoolhouse Lane
Portsmouth, RI 02871
(401) 683-9700
Products: Thin and Thick-Film Resistor Chips and Networks.

IRC WIRE AND FILM TECHNOLOGY (TT ELECTRONICS)
736 Greenway Road
Boone, NC 28607
(828) 264-8861
(828) 264-8865 (Fax)
Products: Chromaxx™ (Thin-Film Resistors), Specialty Resistor Networks.

IWAKI MUSEN KENKYUSHO CO., LTD.
Iwaki building
6-2-3 Kanagawa Prefecture Kawasaki City
Takatsu Ku 213-8566
Japan
(044) 833-4311
Products: Resistor Networks, Wirewound.

JAPAN RESISTOR MFG. CO., LTD.
2315 Kitano, nanto-shi
Toyama 939-1897
Japan
(81-76) 362-1180
Products: Variable Resistors. Also carries a line of Fixed Wirewound, Resistor Networks, and Axial Leaded Metal Film.

KOA CORP.
2-17-2 Midori-Cho
Fuchu-Shi, Tokyo 183-0006
Japan
(81-42) 336-5755
(81-42) 336-5353 (Fax)
Products: Metal Film, Wirewound, Chips (Thick and Thin-Film), Resistor Networks, Variable Resistors, Trimmers, and Potentiometers.

KOA SPEER ELECTRONICS, INC.
199 Bolivar Drive
Bradford, PA 16701
(814) 362-5536
(814) 362-8883 (Fax)
Products: Leading supplier of surface mount components including resistors, inductors, resistor networks, integrated components, ferrite beads, EMI filters and circuit protection components.

MERRIMAC INDUSTRIES, INC.
41 Fairfield Place
West Caldwell, NJ 07006
(973) 575-1300
Products: Resistor Networks.

METALLUX SA
via Moree 12
6850 Mendrisio
Switzerland
(41-091) 640-6450
(41-091) 640-6451 (Fax)
Products: Thin and Thick-Film Resistor Networks.

MINI-SYSTEMS, INC.
20 David Road
North Attleboro, MA 02761
(508) 695-0203
Products: Thick and Thin-Film Resistor Chips & Networks and Custom Hybrid Integrated Circuits.

NIKKOHM CO., LTD.
1-3-6 Nihonbashi-Ningyocho
Chuoku, Tokyo 103-0013
Japan
(81-33) 664-1391
(81-33) 664-5770 (Fax)
Products: Metal Film, Thick and Thin-Film Resistor Networks.

OSI OPTOELECTRONICS AS
PO Box 83
No-3191 Horten
Norway
(47-33) 03-0300
(47-33) 04-9310 (Fax)
Products: Fixed Resistor and Line Feed Surge Networks.

PANASONIC ELECTRONIC DEVICES CO., LTD.
1006 Kadoma
Kadoma City, Osaka 571-8506
Japan
(81-66) 906-1652
Products: MLCCs, MLC Leaded and SLCs.
Panasonic is primarily known as a blanket supplier of both fixed and variable on-board capacitors; they supply as much product by type, configuration and dielectric as they possibly can on a worldwide basis. Panasonic has been widely successful in DC film capacitors and aluminum electrolytic capacitors outside of Japan. They also produce ceramic capacitors and tantalum capacitors for captive consumption and merchant market sales.

QCIRCUITS
(Formerly Bel-Tronics)
2775 Algonquin Road, Suite 300
Rolling Meadows, IL 60008
(630) 543-7777
(630) 543-0038 (Fax)
Products: Thin and Thick-Film Resistor Networks.

RCD COMPONENTS, INC.
520 E. Industrial Park Drive
Manchester, NH 03109
(603) 669-0054
(603) 669-5455 (Fax)
Products: Supplier of Axial and Radial Leaded Through-Hole Resistors for power and related markets.

RIVER ELECTRONICS PTE. LTD.
49 Jalan Pemimpin #04-03
Aps Industrial Building
Singapore 577203
(65-6) 258-7874
(65-6) 258-7366 (Fax)
Products: Carbon Film, Metal Film and Resistor Networks. A Primary Manufacturer of Carbon Film Resistors.

ROHM ELECTRONICS USA, LLC

Product: Ultra miniature Chip Resistor Networks from ROHM

Chip networks significantly reduce the number of components required, saving mounting space as well as increasing cost efficiency. ROHM network resistors are produced with the same high quality and standards as our chip resistors. Convex-type electrodes simplify visual fillet inspection by allowing the use of automatic inspection equipment. ROHM offers a broad lineup of network resistors in a variety of sizes (from 0402 to 1206), element configurations (2-/4-/8-element), and resistances (10Ω-1MΩ).
For more information on these products or to find a ROHM representative in your area, please visit www.rohmelectronics.com.

Marketing76@rohmelectronics.com
10145 Pacific Heights Blvd., Ste. 1000
San Diego, CA 92121
Phone: 888-775-ROHM
www.rohmelectronics.com

RUWIDO AUSTRIA GMBH & CO.
Kostendorter Strasse 8
A5202 Neumarkt
Austria
(43 62) 164-5710
(436) 216-7291 (Fax)
Products: Carbon Film, Thick and Thin-Film Resistor Networks, R/C Networks, and Variable Resistors.

SANSHIN ELECTRIC
16-8, Uchikanda 1-Chome, Chiyoda-Ku
Tokyo 101-0047
Japan
(03-3) 295-1831
(03-5) 259-8041 (Fax)
Products: Variable Resistors and Fixed Resistor Networks.

RESISTORS

STATE-OF-THE-ART, INC.
2470 Fox Hill Road
State College, PA 16803
(814) 355-8004
Products: Thin and Thick-Film Resistor Chips and Networks.
This company is a prime military supplier.

SUSUMU CO., LTD.
14 Umamawashi-Cho
Kamitoba, Minami-Ku
Kyoto, 601-8177
Japan
(81-75) 662-7154
(81-75) 671-7374 (Fax)
Products: Thin and Thick-Film Resistor Networks.

TAI ELECTRONIC CO., LTD.
No.133 Nan-Kann Road Sec 1
Lu-Jwu, Taoyuan
Taiwan
(03) 322-1988
Products: Metal Film, Metal Oxide, Resistor Networks, Flat Chip Resistors, Carbon Film, and Wirewound.

TAIYO DENKI CO., LTD. (TAIYOHM)
690-0021 Shimane prefecture
Matsue City, Yada Cho 250-100
Japan
(085) 224-3674
(085) 227-1280 (Fax)
Products: Carbon Film, Metal Film, Metal Oxide, Wirewound, Resistor Networks, and Flat Resistor Chips.

TALEMA ELECTRONIK GMBH
Sembdnerstr. 5
Postfach 2523, 82110
Germany
(49-8) 984-1000
(49-89) 841-0025
Products: Resistor Networks.

TEIKOKU TSUSHIN KOGYO CO., LTD.
335 Kariyado, Nakahara-ku,
Kawasaki 211-8530
Japan
(81-44) 411-3211
(81-44) 434-3622 (Fax)
Products: Resistor Networks, Metal Film, Metal Oxide, and Variable Resistors.

TEPRO-VAMISTOR
12449 Enterprise Blvd.
Largo, FL 33762
(727) 796-1044
(727) 535-3508 (Fax)
Products: Metal Film, Wirewound, and Resistor Networks.

THIN-FILM TECHNOLOGY CORP. (SUSUMU)
1980 Commerce Drive
North Mankato, Minnesota 56003
(507) 625-8445
(507) 625-3523 (Fax)
Products: Thin and Thick-Film Resistor Networks.

TOKYO COSMOS ELECTRIC CO., LTD.
2-268 Sobudai, Zama
Kanagawa 228-8510
Japan
(046) 253-2111
(046) 253-3640
Products: Resistor Networks, Variable Resistors, and Potentiometers.

TTI, INC.

Product: Never Short on Solutions

Headquartered in Fort Worth, Texas, TTI, Inc. is a distributor specialist of passive, interconnect, electromechanical, and discrete components. TTI is the distributor of choice for industrial and consumer electronic manufacturers worldwide.

TTI's product line includes: resistors, capacitors, connectors, potentiometers, trimmers, magnetic and circuit protection components, wire and cable, wire management, identification products, application tools, electromechanical devices, and discrete components. We distribute these products from a broad line of manufacturers.

TTI strives to be the industry's preferred information source by offering, through the tti|MarketEye blog, the latest IP&E technology and market information, technical seminars, RoHS seminars, industry research reports and much more.

TTI employs more than 2,000 people at more than 50 locations throughout North America, Europe, and Asia.

Sales Contact: information@ttiinc.com
2441 Northeast Pkwy
Fort Worth, Texas 76106
Phone: 800-CALL-TTI (225-5884)
www.ttiinc.com

VISHAY INTERTECHNOLOGY, INC.
63 Lancaster Avenue
Malvern, PA 19355-2143
(610) 644-1300
(610) 296-0657 (Fax)
Products: Foil Resistors; Film Resistors (Metal Film Resistors, Thin Film Resistors, Thick Film Resistors, Metal Oxide Film Resistors, Carbon Film Resistors); Wirewound Resistors; Power Metal Strip® Resistors; Chip Fuses; Variable Resistors (Cermet Variable Resistors, Wirewound Variable Resistors, Conductive Plastic Variable Resistors); Networks/Arrays; Non-Linear Resistors (NTC Thermistors, PTC Thermistors, Varistors).

Resistor Brands:
Vishay Angstrohm
Vishay Beyschlag
Vishay BCcomponents
Vishay Dale
Vishay Draloric
Vishay Electro-Films
Vishay Foil Resistors
Vishay Phoenix
Vishay Sfernice
Vishay Spectrol
Vishay Techno
Vishay Thin Film
and others

WILBRECHT ELECTRONICS, INC.
1400 Energy Park Drive
Suite 18
St. Paul, MN 55108-5248
(651) 659-0919
(651) 659-9204 (Fax)
Products: Nichrome Precision Metal Film Resistors.

YAGEO CORP.
3F, 233-1 Pao Chiao Road
Hsin Tien, Taipei
Taiwan, ROC
(88-622) 917-7555
(88-622) 917-3789 (Fax)
Products: SIP Networks and Arrays.

RESISTORS

Through-Hole Resistor Suppliers

AKAHANE ELECTRONICS
7116 Inabe Ina
Nagano 396-0008
Japan
(81-26) 578-2312
Products: Carbon, Metal Film, and Metal Oxide Leaded Resistors.
Relationship with SEI in the United States.

ALPHA ELECTRONICS CORP.
4th Fl, MG Ikenohata Bldg.
1-2-18 Ikenohata
Taito-ku, Tokyo 110-0008
Japan
(81-3) 5832-6450
(81-3) 5832-6451 (Fax)

CADDOCK ELECTRONICS, INC.

Products: High Performance Film Resistors

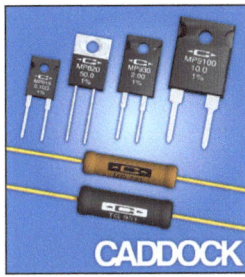

Caddock Electronics develops and manufactures High Performance Film Resistor products:
- TO-Style Power Film Resistors – with power ratings from 15 Watts up to 100 Watts at 25°C case temperature, 1% tolerance on most resistance values.
- Current Sense Resistors – resistance to 0.005 ohm and current ratings up to 60 amps.
- Non-Inductive Power Film Resistors – in axial and radial lead packages.
- Ultra-Precision with Low TC – Radial lead ceramic packages with absolute TC's as low as 2 ppm/°C, tolerances as tight as 0.01%, and resistances from 50 ohms to 25 Megohm.
- High Voltage Resistors: High Precision with low TC – Abs. TC as low as 10 ppm/°C, tolerances as tight as 0.1%, and resistances to 200 Megohms at 15KV.
- High Voltage – voltage ratings up to 48KV and resistance to 10 Gigohms.
- Matched Sets, Selected Resistor Sets, and custom resistor devices.

Applications Engineering
caddock@caddock.com
17271 N Umpqua Hwy
Roseburg OR 97470
Phone: (541) 496-0700
www.caddock.com

EPCOS AG
(EPCOS and TDK will merge in 2009)
P.O. Box 80 17 09
81617 Munich
Germany
(49) 896-3609
(49) 896-362-2689 (Fax)

FUZETEC TECHNOLOGY CO., LTD.
No. 60, Wu-Kung 5 Road
Wu Gu Industrial Park
Taipei Hsein
Taiwan 24890
(88-68) 990-2113
Products: Polymeric PTC (PPTC) resettable fuses.

FUKUSHIMA FUTABA ELECTRIC CO.
12-14, Ikegami, 8-Chome
Ota-Ku, Tokyo 146-0082
Japan
(035) 700-3611
(035) 700-3343 (Fax)
Products: Nichrome Axial Leaded Resistors.

HUNTINGTON ELECTRIC
550 Condit Street
Huntington, IN 46750
(260) 356-0756
Products: Axial lead resistors, fixed and adjustable products, oval resistors, edge wound styles, printed circuit board- and capacitor-mounted resistors, and products for military application.

IWAKI MUSEN KENKYUSHO CO., LTD.
Iwaki building
6-2-3 Kanagawa prefecture Kawasaki city
Takatsu Ku 213-8566
Japan
(044) 833-4311
(044) 833-6605
Products: Wirewound Resistors.

KAMAYA ELECTRIC CO., LTD. (WALSIN)
A-209 KSP R&D Business Park Building
3-2-1 Sakado, Takatsu-ku, Kawasaki-shi, Kanagawa 213-0012
Japan
(81-44) 820-8560
(81-44) 820-8563 (Fax)
Products: Axial Leaded Resistors for high voltage and surge applications in industrial electronics.

KOA CORP.
2-17-2 Midori-Cho
Fuchu-Shi
Tokyo 183-0006
Japan
(81-42) 336-5755
(81-42) 336-5353 (Fax)
Products: Flat Chip Resistors and Arrays, Metal Film and Carbon Film Leaded Resistors.

KOA SPEER ELECTRONICS, INC.
199 Bolivar Drive
Bradford, PA 16701
(814) 362-5536
(814) 362-8883 (Fax)
Products: Leading supplier of surface mount components including resistors, inductors, resistor networks, integrated components, ferrite beads, EMI filters and circuit protection components.

OHMITE MANUFACTURING CO., INC.
1600 Golf Rd., Suite 850
Rolling Meadows, IL 60008
(866) 964-6483
(847) 574-7522
Products: Wirewound and Axial Leaded Power Film Resistors.

PANASONIC ELECTRONIC DEVICES CO., LTD.
1006 Kadoma, Kadoma City
Osaka 571-8506
Japan
(81-66) 906-1652
Products: MLCCs, MLC Leaded and SLCs.
Panasonic is primarily known as a blanket supplier of both fixed and variable on-board capacitors; they supply as much product by type, configuration and dielectric as they possibly can on a worldwide basis. Panasonic has been widely successful in DC film capacitors and aluminum electrolytic capacitors outside of Japan. They also produce ceramic capacitors and tantalum capacitors for captive consumption and merchant market sales.

RCD COMPONENTS, INC.
520 E. Industrial Park Drive
Manchester, NH 03109
(603) 669-0054
(603) 669-5455 (Fax)
Products: Supplier of Axial and Radial Leaded Through-Hole Resistors for power and related markets.

RFE INTERNATIONAL, INC.
17691 Mitchell North, Unit B
Irvine, CA 92614
(949) 833-1988
(949) 833-1788 (Fax)
Products: Capacitors, Diodes/Rectifiers, Inductors, MOVs PPTCs, Thermistors, Resistors.

ROHM CO., LTD.
21 Saiin Mizosaki-cho
Ukyo-ku, Kyoto 615
Japan
(81-75) 311-2121
Products: Thick-Film Chip Resistors and Arrays, Metal Film, and Metal Oxide Leaded Resistors.

SRT RESISTOR TECHNOLOGY GMBH
Ostlandstr. 31
90556 Cadolzburg
Germany
(49-09) 130-7952
Products: Special Chip & Leaded Resistors for High Value, High Voltage, and High Precision Applications.

STACKPOLE ELECTRONICS, INC
2700 Wycliff Road
Raleigh, NC 27607
(919) 850-9500
(919) 850-9504 (Fax)
Products: Supplier of Wirewound and Metal Film Resistors. This company is related to Akahane in Japan.

RESISTORS

TTI, INC.

Product: Never Short on Solutions

Headquartered in Fort Worth, Texas, TTI, Inc. is a distributor specialist of passive, interconnect, electromechanical, and discrete components. TTI is the distributor of choice for industrial and consumer electronic manufacturers worldwide.

TTI's product line includes: resistors, capacitors, connectors, potentiometers, trimmers, magnetic and circuit protection components, wire and cable, wire management, identification products, application tools, electromechanical devices, and discrete components. We distribute these products from a broad line of manufacturers.

TTI strives to be the industry's preferred information source by offering, through the tti|MarketEye blog, the latest IP&E technology and market information, technical seminars, RoHS seminars, industry research reports and much more.

TTI employs more than 2,000 people at more than 50 locations throughout North America, Europe, and Asia.

Sales Contact: information@ttiinc.com
2441 Northeast Pkwy
Fort Worth, Texas 76106
Phone: 800-CALL-TTI (225-5884)
www.ttiinc.com

VITROHM GMBH (YAGEO)
Ramskamp 70 D
25337 Elmshorn
Germany
(490-412) 187-0103
(490-412) 187-0289 (Fax)
Products: Leaded and Surface-Mount Resistors based upon Nichrome Metal Film.

WILBRECHT ELECTRONICS, INC.
1400 Energy Park Drive
Suite 18
St. Paul, MN 55108-5248
(651) 659-0919
(651) 659-9204 (Fax)
Products: Nichrome Precision Metal Film Resistors.

WORLD PRODUCTS INC.
19654 Eighth Street East
Sonoma, CA 95476
(770) 996-5201
Products: Thermal Fuse, Metal Oxide, Multi-Layer, and Silicon Varistors.

VISHAY INTERTECHNOLOGY, INC.
63 Lancaster Avenue
Malvern, PA 19355-2143
(610) 644-1300
(610) 296-0657 (Fax)
Products: Foil Resistors; Film Resistors (Metal Film Resistors, Thin Film Resistors, Thick Film Resistors, Metal Oxide Film Resistors, Carbon Film Resistors); Wirewound Resistors; Power Metal Strip® Resistors; Chip Fuses; Variable Resistors (Cermet Variable Resistors, Wirewound Variable Resistors, Conductive Plastic Variable Resistors); Networks/Arrays; Non-Linear Resistors (NTC Thermistors, PTC Thermistors, Varistors).

Resistor Brands:
Vishay Angstrohm
Vishay Beyschlag
Vishay BCcomponents
Vishay Dale
Vishay Draloric
Vishay Electro-Films
Vishay Foil Resistors
Vishay Phoenix
Vishay Sfernice
Vishay Spectrol
Vishay Techno
Vishay Thin Film
and others

THE PAUMANOK GROUP

Gas Discharge Tube Arresters: World Markets, Technologies & Opportunities: 2008—2013

This newly released market research report is an 88-page detailed analysis of the global market for gas discharge tube arrester markets, technologies, and opportunities. The study focuses on the global value and volume of demand for surface mount, two-terminal, and three-terminal gas tube arresters. Included in the study is a detailed analysis of competition by gas discharge tube type and form factor by case size.

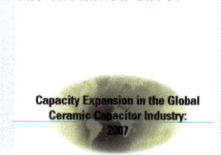

The study also includes consumption by end-use market segment, with a detailed analysis of competition by end-use market segment in telecommunications infrastructure, AC power line markets, and specialty electronic sub-assemblies, including applications in central office and station class infrastructure markets, CATV systems, customer premises equipment and branch line protection, and line voltage equipment protection markets.

A vertical integration market analysis of consumption of gas tubes in SPDs is also given, including applications in coaxial cable connectors, station class protectors, DIN Rail (NPE protectors), and panel-mount and wall-mount protectors. Sales and market shares for all major manufacturers are included, as is gas tube vendors by major distributor. Also included is an analysis of competitive technology as well as a look at synergistic components.

Price: $1250 USD | **Additional Copies:** $125 USD | **Downloadable (PDF) Copy:** $1250 USD
Product Code: GDT2008
Web: www.paumanokgroup.com/market_reports/reports.asp?c=4&p=2

For more details on these reports and others, visit: www.paumanokgroup.com

INDUCTORS

Chip Inductor, Ferrite Bead, & Array

ABC TAIWAN ELECTRONICS CORP.
No. 422, Sec. 1, Yang-Fu Rd.
Yangmei 326 Taoyuan
Taiwan
(88-63) 478-8188
(88-63) 475-5503 (Fax)
Products: Wound Chip Inductor, Power Inductor, Line Filter, Transformer, and Through Hole.

ABCO ELECTRONICS CO., LTD.
5448-4, Sangdaewon-dong
Jungwon-gu
Seongnam, Kyeongkido
Korea
(82-31) 730-5000
(82-31) 743-2824 (Fax)
Products: SMD, EMI Filter, Wire Wound Chip Inductor, Axial & Radial Type Inductor.

BOURNS, INC.
1200 Columbia Ave.
Riverside, CA 92507
(951) 781-5690
Products: 2100HT, 2200HT, and 2300HT High Temperature Toroid Power Industor Series.

CERATECH CORP.
175-3, Guemjeong-Dong, Gunpo-Si
Gyeonggi-Do 435-824
Korea
(82-31) 458-1010
Products: Ceramic Inductors/Ferrite Beads/Chip Varistor.

CHILISIN ELECTRONICS CORP. (YAGEO)
No. 29, Lane 301, Tehhsin Rd.
Hosin, Hukou, Hsinchu
Taiwan
(88-63) 599-2646
Products: Chip beads, chip inductors, power inductors, common mode filters, high current CPU choke and LTCC RF components.

COILCRAFT, INC.
1102 Silver Lake Road
Cary, IL 60013
(800) 322-2645
(847) 639-6400
(847) 639-1469 (Fax)
Products: Chip and Bead Inductors; especially Wirewound Chip Inductors.

EPCOS AG
(EPCOS and TDK will merge in 2009)
P.O. Box 80 17 09
81617 Munich
Germany
(49) 896-3609
(49) 896-362-2689 (Fax)

KEMET CORPORATION

Product: Inductors & Ferrite Beads

KEMET offers a full line of chip inductors and ferrite beads. This includes a line of multi-layer and wirewound surface mount inductors. Products include shielded power inductors for DC/DC converters, chip inductors for conductive noise suppression in power and signal lines, as well as RF inductors for frequency matching.

In addition, KEMET offers a high quality line of ferrite beads to combat radiant noise in both power lines and signal lines. Offered in both standard and high current versions, these are available in case sizes as small as 0201. They are excellent solutions for EMI issues.

With KEMET, you simply won't find an electronic components manufacturer more passionate about customer service, more determined to find new technological solutions to customer problems, and more committed to product quality and on-time delivery. It's how we've helped customers succeed for more than 85 years.

Scott Carson
Product Manager
Phone: (864) 228-4065
www.kemet.com

KOA CORP.
2-17-2 Midori-Cho
Fuchu-Shi
Tokyo 183-0006
Japan
(81-42) 336-5755
(81-42) 336-5353 (Fax)
Products: Flat Chip Resistors and Arrays, Metal Film and Carbon Film Leaded Resistors.

KOA SPEER ELECTRONICS, INC.
199 Bolivar Drive
Bradford, PA 16701
(814) 362-5536
(814) 362-8883 (Fax)
Products: Surface mount components including resistors, inductors, resistor networks, integrated components, ferrite beads, EMI filters, and circuit protection components.

MURATA ELECTRONICS, N.A., INC.

Product: Chip Inductor, Ferrite Bead and Array

Murata Electronics offers an extensive line of magnetic products that includes but is not limited to Ferrite Beads, Ferrite Bead Arrays and Inductors. Murata Electronics' Ferrite Beads represented by the BLM series are available in a wide variety of sizes and electrical characteristics. With sizes as small as EIA 01005 (0.4mm x 0.2mm) and improved features such as low DC resistance, increased current capacity and high frequency filtering, the BLM series is suitable for a broad range of applications. In addition to the BLM series, Murata Electronics also offer Ferrite Bead Arrays, the new BLA2A series. This product is capable of filtering 4 lines in a compact 2.0mm x 1.0mm package making it ideal for limited board space applications. As with the Ferrite Beads, Murata Electronics is also a leading innovator in surface mount inductors. Utilizing three manufacturing technologies, Multilayer, Thin Film and Wirewound technologies, Murata offers a diverse line of inductors suitable for both RF and Power applications.

Deryl Kimbro
Group Product Manager
dkimbro@murata.com
2200 Lake Park Drive
Smyrna, GA 30080
Phone: (770) 436-1300
www.murata.com

MURATA MANUFACTURING CO.
10-1, Higashikotari 1-chome
Nagaokakyo-shi, Kyoto 617-8555
Japan
(81-75) 951-9111
Products: Chip Inductors.

Continued on page 41

THE PAUMANOK GROUP

Get the latest in passive component market research from Paumanok Publications

NEW!

Passive Electronic Components: World Market Outlook: 2009–2014

This report was designed for end-users of passive electronic components who need to update their supply chain information for passive parts on an annual basis. This report was designed with input from major purchasers of passive electronic components and in direct response to industry needs.

Which Components Are Tracked In This Report?

This report forecasts global consumption volume, value and pricing for capacitors, resistors and inductors, including MLCC, Aluminum, Tantalum, DC Film and AC Film Capacitors; Thick Film Chip Resistors, Resistor Networks, Integrated Passive Devices; Metal Film Resistors, Tin-Oxide Resistors, Wirewound Resistors, Carbon Film Resistors; Multilayered Chip Inductors, Ferrite Beads and Bead Arrays and Wirewound Inductor Coils.

Who Is Discussed In This Report?

All major vendors of passive components are discussed in this report, including:

AVX/Kyocera Corporation (NYSE: AVX and RIC: 6971)
EPCOS AG (NYSE: EPC)
KEMET Electronics (NYSE: KEM)
KOA Corporation (RIC: 6999)
Nichicon Corporation (RIC: 6996)
Nippon Chemi-Con Corporation (RIC: 6997)
Panasonic Electronic Devices Company Limited (RIC: 6752)
Murata Manufacturing Limited (RIC: 6981)
Rohm Company Limited (RIC: 6963)
Rubycon Corporation (Private)
Samsung Electro-Mechanical (KSE: 009150)
Sumida Electric Company Limited (RIC: 6817)
Taiyo Yuden Corporation (RIC: 6976)
TOKO Inc. (RIC: 6801)
Tokin Corporation/NEC (RIC: 6795)
Vishay Intertechnology, Inc. (NYSE: VSH)
Walsin Technology (TT: 5335)
Yageo Corporation (TW: 2327)

What Are Some of The Major Conclusions Of The Report?

The report makes future price, value, and volume forecasts for each passive component. A current analysis of each end use market segment, including:

- wireless handsets, personal computers, television sets
- automotive electronics
- large home appliances
- power supplies
- infrastructure
- lighting
- defense
- medical

These end-use market segments are used to make short-term and long-term assumptions about each of the key passive electronic components. This report was released in February 2009 and documents the impact of the global economic downturn on the passive component industry and makes forecasts for its recovery to 2014. 199 Pages.

Price: $2500 USD | **Additional Copies:** $250 USD
Downloadable (PDF) Copy: $2500 USD
Product Code: CRL2009
Web: www.paumanokgroup.com/market_reports/ppf/c/1/reports.asp

For more details on these reports and others, visit:
www.paumanokgroup.com

INDUCTORS

Continued from page 39

NIC COMPONENTS CORP.

Products: Inductors, Ferrite Beads, and Arrays

NIC Components has expanded their offering of surface mount magnetics to include both shielded and non-shielded type surface mount power inductors. The new NPI series of inductors from NIC are available in power ratings from 2-70A. It includes miniature and low profile types for hand held and height restrictive applications. NIC's new Quick Kit program allows designers to create a customized inductor kit from over 400 available sizes and styles with a few mouse clicks on the www.nic-comp.com or www.smtmagnetics.com websites. In addition, NIC offers a line of ferrite beads, multilayer, wire wound, and thin film inductors. NIC Components has direct sales offices and warehousing facilities in the US, Europe, and Asia to provide sales, logistics, and technical support to our customers worldwide.

Eric Moller
National Sales Manager
sales@niccomp.com
70 Maxess Road,
Melville, NY 11747
Phone: (631) 396-7500
www.niccomp.com

PANASONIC ELECTRONIC DEVICES CO., LTD.
1006 Kadoma, Kadoma City
Osaka 571-8506
Japan
(81-66) 908-1211
Products: MLCCs, MLC Leaded and SLCs.
Panasonic is primarily known as a blanket supplier of both fixed and variable on-board capacitors; they supply as much product by type, configuration and dielectric as they possibly can on a worldwide basis. Panasonic has been widely successful in DC film capacitors and aluminum electrolytic capacitors outside of Japan. They also produce ceramic capacitors and tantalum capacitors for captive consumption and merchant market sales.

RFE INTERNATIONAL, INC.
17691 Mitchell North, Unit B
Irvine, CA 92614
(949) 833-1988
(949) 833-1788 (Fax)
Products: Capacitors, Diodes/Rectifiers, Inductors, MOVs PPTCs, Thermistors, Resistors.

STEWARD
A Unit of Laird Technologies
1200 E. 36th Street
Chattanooga, TN 37407
(423) 308-1690
Products: Ferrite EMI chip beads.

SUMIDA ELECTRIC CO., LTD.
Asahi Building, 3-12-2, Nihonbashi
Chuo-ku, Tokyo 103-8259
Japan
(81-03-5) 202-7111
(81-03-5) 202-7104 (Fax)
Products: DC/DC Converter, EMC, IFT/RF Coils, Telecommunication, Antenna Coils, RFID, LC Filter, D-Amp Filter, EL Drive Inductor, Multiple Power Inductor, Power Choke Inductor, Inverter Module, Inverter Transformer, ABS Coil, ABD Module, Actuator.

TAIYO YUDEN (USA), INC.

Product: NR6045T Series Power Inductors

Taiyo Yuden (U.S.A.) Inc., offers a complete line of 3-, 4-, 6- and 8 mm-square wire-wound chip inductors. The new 6 mm² NR6045T series provides a 4.5 mm maximum height for DC-DC converter choke coil applications in LCD / plasma TVs and other flat panel displays (FPDs). The 14 models in the NR6045T series offer an excellent choice of low DC resistance (0.014 to 0.5 Ohms), high inductance (1.0 to 100 µH) and high current ratings (0.8 to 8.5 A). The NR6045T-100M device, for example, offers the industry's highest rated current (3.0 A @ 10 µH) in the 6 mm² form factor. With the equivalent performance of larger 7 mm² to 10 mm² devices, the 26% to 64% smaller footprint, respectively, of NR6045T series devices facilitates more compact DC-DC converters and thinner-profile product designs. For further information, visit Taiyo Yuden (U.S.A.) Inc. at www.yuden.us.

Yaeko Minamikawa
Marketing Analyst
sales@t-yuden.com
1930 Thoreau Dr., Ste 190
Schaumburg, IL 60173
Phone: (847) 925-0888
www.yuden.us

INDUCTORS

TDK CORP.
(TDK and EPCOS will merge in 2009)
1-13-1 Nihonbashi
Chuo-ku, Tokyo 103-8272
Japan
(81-33) 278-5111
Products: Producer of Chip and Bead Inductors.

TOKO, INC.
1-17, Higashi-Yukigaya 2-chome
Ohta-ku, Tokyo, 145-8585
Japan
(81-33) 727-1166
(81-33) 727-1653 (Fax)
Products: Coils / Inductors and LC Filters, Common Mode Choke Coils and Interface Modules, Ceramic Filters and Resonators, Dielectric Filters and Antennas, Multilayer Products (Chip Inductors, Filters, Diplexers, Etc.).

TTI, INC.

Product: Never Short on Solutions

Headquartered in Fort Worth, Texas, TTI, Inc. is a distributor specialist of passive, interconnect, electromechanical, and discrete components. TTI is the distributor of choice for industrial and consumer electronic manufacturers worldwide.

TTI's product line includes: resistors, capacitors, connectors, potentiometers, trimmers, magnetic and circuit protection components, wire and cable, wire management, identification products, application tools, electromechanical devices, and discrete components. We distribute these products from a broad line of manufacturers.

TTI strives to be the industry's preferred information source by offering, through the tti|MarketEye blog, the latest IP&E technology and market information, technical seminars, RoHS seminars, industry research reports and much more.

TTI employs more than 2,000 people at more than 50 locations throughout North America, Europe, and Asia.

Sales Contact: information@ttiinc.com
2441 Northeast Pkwy
Fort Worth, Texas 76106
Phone: 800-CALL-TTI (225-5884)
www.ttiinc.com

VISHAY INTERTECHNOLOGY, INC.
63 Lancaster Avenue
Malvern, PA 19355-2143
(610) 644-1300
(610) 296-0657 (Fax)
Products: : Power Inductors (SMD and Through-Hole, Molded, Ferrite, Toroidal); RF Inductors (SMD and Through-Hole, Wirewound, Multilayer, Film); Custom Inductors and Transformers (SMD and Through-Hole); Connectors (Edgeboard, Rack and Panel, High-Reliability Custom); Frequency Control (SMD and Through-Hole Crystals and Oscillators).

Inductor Brands:
Vishay Dale

WALSIN TECHNOLOGY CORP.
566-1, Kao-Shi Road, Yang-Mei
Tao-Yuan
Taiwan
(88-63) 475-8711
(88-63) 475-7130 (Fax)
Products: Ferrite Beads

WORLD PRODUCTS INC.
19654 Eighth Street East
Sonoma, CA 95476
(770) 996-5201
Products: Common Mode, EMI and LC Filters; Ceramic, Ferrite, and Micro Inductors; Ferrite Beads and Arrays; ISOCOUPLER-Photo; ISMOS Photo MOS Relays; TVS Diodes; Thyristors; Surge Absorbers.

INDUCTORS

Ferrite Core Suppliers

AVX/TPC
Av. Du Colonel Prat.
F-21850 St. Appollinaire
France
(33) 871-7400
Products: Chip Inductors.

EPCOS AG
(EPCOS and TDK will merge in 2009)
P.O. Box 80 17 09
81617 Munich
Germany
(49) 896-3609
(49) 896-362-2689 (Fax)
Products: New Planar Cores for DC/DC Converters.

FERROXCUBE (YAGEO)
Yageo Corporation
3F No. 233-1 Pao-Chiao Rd
Hsin Tien, Taipei
Taiwan, R.O.C.
(88-62) 917-7555
(88-62) 917-3789 (Fax)
Products: Ferrite Cores.
Ferroxcube is part of Yageo's capacitor and resistor operations.

KOA SPEER ELECTRONICS, INC.
199 Bolivar Drive
Bradford, PA 16701
(814) 362-5536
(814) 362-8883 (Fax)
Products: Leading supplier of surface mount components including resistors, inductors, resistor networks, integrated components, ferrite beads, EMI filters and circuit protection components.

NEOSID AUSTRALIA PTY, LTD. (TT ELECTRONICS)
23-25 Percival St.
Lilyfield
New South Wales 2040
Australia
(61-29) 660-4566
(61-29) 552-1748 (Fax)

SAMWHA ELECTRONICS
Seoul HQ
Samyoung Bldg.
587-8 Sinsa-dong
Kangnam-gu, Seoul
Korea
(02) 546-0999
(02) 546-7354 (Fax)
Products: Samwha is known for its capacitor products, but they are also a major supplier of ferrite cores.

TDK CORP.
(TDK and EPCOS will merge in 2009)
1-13-1 Nihonbashi
Chuo-ku, Tokyo 103-8272
Japan
(81-33) 278-5111
Products: Supplier of Ferrite Cores.

TT ELECTRONICS
Clive House
12-18 Queens Road
Weybridge, Surrey
KT13 9XB
United Kingdom
(93) 284-1310
(93) 283-6450 (Fax)
Products: Producer of Advanced Ferrite Cores.
TT Electronics produces advanced ferrite cores through its MMG Division and sister company, Neosid.

TTI, INC.

Product: Never Short on Solutions

Headquartered in Fort Worth, Texas, TTI, Inc. is a distributor specialist of passive, interconnect, electromechanical, and discrete components. TTI is the distributor of choice for industrial and consumer electronic manufacturers worldwide.

TTI's product line includes: resistors, capacitors, connectors, potentiometers, trimmers, magnetic and circuit protection components, wire and cable, wire management, identification products, application tools, electromechanical devices, and discrete components. We distribute these products from a broad line of manufacturers.

TTI strives to be the industry's preferred information source by offering, through the tti|MarketEye blog, the latest IP&E technology and market information, technical seminars, RoHS seminars, industry research reports and much more.

TTI employs more than 2,000 people at more than 50 locations throughout North America, Europe, and Asia.

Sales Contact: information@ttiinc.com
2441 Northeast Pkwy
Fort Worth, Texas 76106
Phone: 800-CALL-TTI (225-5884)
www.ttiinc.com

RAW MATERIAL

Raw Material Suppliers to the Ceramic Capacitor Industry

BASF CORP.
100 Campus Drive
Florham Park, New Jersey 07932
(800) 526-1072

CABOT CORP.
Cabot Supermetals
County Line Rd
P.O. Box 1608
Boyertown, PA 19512-1608
(610) 367-1500
(610) 367-6383 (Fax)

CEPC
3494 Ashby Street
St-Laurent, Québec H4R 2C1
Canada
H4R 2C1
(514) 336-2888
(514) 336-5059 (Fax)
Products: Nickel Electrode Powders for MLCC Electrode.

CERMET MATERIALS, INC.
6 Meco Drive
Wilmington, DE 19804
(302) 999-1447
(302) 999-7211 (Fax)
Products: Processor of Bullion to Co-Precipitated Electrode Powders.

DAIKIN INDUSTRIES, LTD.
Chemical Division
Umeda Center Bldg.
2-4-12, Nakazaki-Nisi, Kitaku, Osaka 530-8323
Japan
(81-66) 373-4312
Products: Precious Metal Electrode Powders and Pastes.

DUPONT ELECTRONICS, INC.
Dupont Microcircuit Materials Division
14 T.W. Alexander Dr.
RTP, NC 27709
(800) 284-3382
Products: Electrode Inks.

FERRO ELECTRONIC MATERIAL SYSTEMS

Product: Advanced Material Systems for MLCC

Ferro Electronic Material Systems is one of the world's largest suppliers of ceramic dielectric powders, inner electrode pastes, termination pastes and binder systems. In addition to offering a vast line of solid-state and chemically precipitated Barium Titanates, Ferro's extensive dielectric formulations for ceramic capacitors include products compatible with precious metal and base metal electrode systems. Our pastes for electrodes and terminations are specifically designed for compatibility with the leading equipment used in the industry to maximize performance with a wide range of dielectric systems. Ferro is unique in offering organic binder materials and systems designed for coating and casting dielectric tapes and other film products.
To provide added value to our customers, Ferro has established R&D and Applied Technology labs in the United States, Europe and Asia to optimize our products' properties and their performance to meet our customers' manufacturing process and product specifications.

Customer Service
mlcc@ferro.com
Ferro Corporation
1000 Lakeside Ave.
Cleveland, OH 49114
Phone: (216) 641-8580
www.ferro.com

FUJI TITANIUM INDUSTRY CO., LTD.
3-6-32 Nakanoshima (Dai Bldg.)
Kita-ku, Osaka 530-6591
JAPAN
(81-66) 441-6856
(81-66) 441-6855 (Fax)
Products: Titanate Materials.

GWENT ELECTRONIC MATERIALS, LTD.
Monmouth House, Mamhilad Park
Pontypool, Gwent NP4 0HZ
United Kingdom
(44-149) 575-0505
Products: MLC Electrode and Termination Materials.

HAIKU TECH, INC.

Products: Manufacturing and Materials Solutions for Microelectronic Components

Haiku Tech specializes in innovative materials, equipment and technology to efficiently produce passive components like MLCC's, MLV, MLI, LTCC, piezo actuators, Solid Oxide Fuel Cells (SOFC) and SOLAR Cells.

Our products include:
- Dielectric BME and PME Compositions, Microwave dielectrics, High purity Barium Titanates
- Water Tape Binders, Paste vehicle systems for copper paste, Specialty pastes
- Tape casters, Punching machines, Screen Printers, automated multi-layer stacking/collating lines, Automated Screen printing lines, Lamination Presses, Cutters
- Chip Metallization (new for 0201 and arrays) and tooling
- In-line thickness gauges
- Electroplating lines
- Automatic Vision Inspection and Sorting Equipment
- High Temperature Optical Dilatometers and Microscopes
- High Temperature vision analysis of materials

Mr. Martin De Moya (America)
Sales and Service Manager
sales@haikutech.com
1669 NW 79th Avenue
Miami, FL 33126, USA
Phone: (305) 463-9304

Roderik Höppener (Europe)
President and CEO
Dorpsstr 100A
6274 NN Reijmerstok
The Netherlands
Phone: +31 43 4578054
www.haikutech.com

RAW MATERIAL

H.C. STARCK GMBH
Im Schleeke 78-91
38642 Goslar
Germany
(49-5) 321-7510
Products: Niobate Additives.

Other Plant Locations:

H.C. STARCK INC.
45 Industrial Place
Newton, MA 02161-1951
(617) 630-5800
Products: Niobate Additives.

H.C. STARCK LTD.
1-30-5 Hamamatsucho
Minato-ku
Tokyo 105-0013
Japan
(81-35) 776-5001
Products: Niobate Additives.

H.C. STARCK CO., LTD.
5, I-3A Road, Map Ta Phut
Industrial Estate
Rayong 21150
Thailand
(66-38) 683-077

HERAEUS CMD
Thick Film Division
24 Union Hill Road
West Conshohocken, PA 19428
(610) 825-6050
Products: MLCC Metallization Materials.

INCO SPECIAL PRODUCTS
2101 Hadwen Road
Mississauga, Ontario
Canada L5K2L3
(905) 403-3350
Products: Nickel Electrode Powders.

JFE MINERAL CO., LTD.
5th Floor, Shibakoen First Building
8-2 Shiba 3-chome, Minato-ku
Tokyo 105-0014
Japan
(03-4) 455-2210
(03-4) 455-2266 (Fax)
Products: Capacitor Nickel Electrode Powder.

JOHNSON MATTHEY
40-42 Hatton Garden
London EC1N 8EE
(0-20-7) 269-8400
(44-0-2-07) 269-8433 (Fax)
Products: World Supplier of Component Materials, Including Precious Metal Chlorides, Nitrates, Oxides, Acetates, and Complex Inorganic Salts. Most compounds are available in bulk quantities to customized specifications.

KCM CORP.
2-41, Tsukisan-cho
Minato-ku Nagoya
AICHI, 455-8668
Japan
(81-52) 661-3180
Products: Titanate Materials.

LYDALL-SOLUTECH B.V.
Eisterweg 4
6422 PN Heerlen
The Netherlands
(31-45) 543-5212
(31-45) 542-7929 (Fax)

2009 EDITORIAL CALENDAR

It's Time to Get Connected to the Passive Component Industry

May/June Issue
New Products, Markets & Opportunities • Deadline: April 30

July/August Issue
Raw Material • Deadline: June 26

September/October Issue
Aluminum/Film Capacitors • Deadline: August 21

November/December Issue
Linear/Non-Linear Resistors & Inductors • Deadline: October 23

RAW MATERIAL

M.E. SCHUPP INDUSTRIEKERAMIK GMBH & CO. KG
Neuhausstraße 4 – 10
52078 Aachen
Germany
(49-024) 193-6770
Products: Ceramic and Metallic Key Components.

METALOR TECHNOLOGIES USA
52 Gardner St.
Attleboro, MA 02703
(508) 226-4470
(508) 695-4180 (Fax)
Products: CPXX04 Series Co-Precipitated Ag/Pd Electrode Powders.

MRA LABORATORIES, INC.

Product: Formulated Ceramic Dielectric and Electrode Materials for the MLCC Industry

MRA Laboratories, Inc. is a leading supplier of formulated ceramic dielectric materials and customized technology solutions to the MLCC industry. We offer a complete line of state-of-the-art, air-fired dielectric formulations. These include low K, mid K, and high K COG, and X7R/X8R dielectrics with dielectric constants up to 4500.
Our VLF (very low-fired) series of dielectrics are compatible with up to 100% Ag electrode systems. In addition, many are "environmentally friendly", as they are not formulated with lead, cadmium, or hexavalent chromium.
MRA's line of formulated dielectrics provides an excellent choice for today's demanding high-frequency and high voltage specialty capacitor applications. Each is compatible with a wide range of binder systems and dispersants. Custom shrinkage-matched electrode inks are available for order with each of our formulated dielectric products.
Now offering RoHS Testing Services to the electronics industry!!

mra@mralabs.com
15 Print Works Drive
Adams, MA 01220
Phone: (413) 743-3927
Fax: (413) 743-0305
www.mralabs.com

NAMICS

Products: HIMEC and UNIMEC Silver and Copper Conductive Paste

Namics manufactures and sells a variety of silver and copper conductive paste for MLCCs, chip resistors, chip inductors, thermistors and other passive components. Insulating materials are also provided for passive components. Encapsulants, such as underfills and globtops, are offered for the semiconductor industry.

sales@namics-usa.com
2055 Gateway Place, Suite 480
San Jose, CA 95110
Phone: (408) 516-4611
www.namics.co.jp

NANODYNAMICS, INC.
901 Fuhrmann Boulevard
Buffalo, NY 14203
(716) 853-4900
(716) 853-8996
Products: ND Copper, ND Nickel, and ND Silver in Micron and Nanometer Sizes.

NIPPON CHEMICAL INDUSTRIAL CO.
11-1, 9-chome
Kameido, Koto-ku, Tokyo 136-8515
Japan
(81-33) 636-8111
(81-33) 636-6817 (Fax)
Products: Barium Carbonate and Titanate Materials.

NYACOL NANO TECHNOLOGIES, INC.

Products: Colloidal Metal Oxides

Nyacol Nano Technologies, Inc. offers a broad line of colloidal metal oxides for use in conductive pastes for MLCC, HTCC, and LTCC components. The 40-80nm oxides are dispersed in mineral spirits for easy introduction into most conductive paste formulations. These colloidal metal oxide dispersions are completely agglomerate free.
Nyacol has been manufacturing this product line for over a decade and can offer single lots in metric ton quantities. For MLCCs, the BtMin has gained wide acceptance as a sintering control agent and shrinkage modifier when used in Ni base metal electrode pastes. ZrMin has been successfully utilized with precious metal pastes. A partial list of these colloidal oxide systems include $BaTiO_3$, ZrO_2, Y_2O_3, Y_2O_3-ZrO_2, ZnO, and MgO. Visit www.nyacol.com to view a complete list of products and TEMs illustrating the high degree of de-agglomeration.

Bob Nehring
rnehring@nyacol.com
Phone: (800) 438-7657 x242
www.nyacol.com

Ken Magrini
Specific Solutions
specificsolutions@verizon.net
Phone: (951) 296-1035

PROTAVIC AMERICA, INC.
8 Ricker Ave.
Londonderry, NH 03053
(800) 807-2294
(978) 372-2016 (Int'l)
Products: Protavic flexible polymer termination.

ROHM & HAAS, CO.
100 Independence Mall West
Philadelphia, PA 19106-2399
(877) 288-5881
Products: Ceramic Binder Materials.

RAW MATERIAL

SACHEM, INC.

Product: Quaternary Ammonium Hydroxides

SACHEM Inc. offers its complete product line of organic strong bases and quaternary ammonium hydroxides (QAOH) such as tetramethylammonium hydroxide (TMAH) and tetrabutylammonium hydroxide (TBAH) for the production of nano-ceramic powders. Other processing aids in the product line include bis-Tetramethylammonium Oxalate. The QAOH acts as a mineralizer and peptizer in chemical preparation methods such as co-precipitation, hydrothermal synthesis, and the sol-gel route. Organic strong bases can improve these processes through reducing impurities, improving stoichiometry, and minimizing aggregation and agglomeration. Recent studies indicate QAOH can aid in controlling nano-crystalline growth (including size, distribution, and morphology), can be easily decomposed at low temperatures, and impart lower viscosities than higher molecular weight surfactants. SACHEM's expertise in creation and commercialization of new QAOH can turn your ideas into reality. Products are available in volumes of 1-gal to IBC. Purity as required. Samples are available.

John Mattson
Market Manager, Asia & Americas
info@sacheminc.com
821 Woodward St.
Austin, TX 78704
Phone: (512) 421-4900
www.sacheminc.com

SAKAI CHEMICAL CO.
1-1-23, Ebisuno-cho-nishi, Sakai,
Osaka 590-8502
Japan
(72) 223-4111
(72) 223-8355 (Fax)
Products: Hydrothermal Barium Titanate.

SAMSUNG FINE CHEMICALS
22nd Fl., Seocho Tower Samsung Life Insurance
1321-15, Seocho-2(i)dong
Seocho-gu, Seoul
Korea
(82-2) 772-1754
(82-2) 772-1779 (Fax)
Products: Barium Titanate Powder.

SHOEI CHEMICAL, INC.
Shinjuku Mitsui Building
1-1, Nishi-Shinjuku 2-chome
Shinjuku-ku, Tokyo 163-0443
Japan
(81-03) 344-6661
Products: Electrode Powders and Pastes. Intermediate supplier of palladium and palladium-silver paste.

SUMITOMO METAL MINING CO., LTD.
11-3, Shimbashi 5-chome
Minato-ku
Tokyo 105 8716
Japan
Products: Electrode and termination Powders and Pastes, including both Precious and Base Metals.

TECHNIC, INC.
Advanced Technology Division
111 E. Ames Court
Plainview, New York 11803
(516) 349-0700
(516) 349-0666 (Fax)
Products: Raw Materials for Ceramic Capacitors.

TORPEDO SPECIALTY WIRE, INC.

Product: Electronic Lead Wires

Torpedo Specialty Wire is a leading manufacturer of electronic lead wires for the capacitor and resistor industry for more than 34 years. The company has led the way in developing new lead wire products and processes to meet the customer's needs. Product line consists of a variety of coatings including Tin/Lead Solders of all compositions, 100% Tin-RoHS compliant, and Nickel with the capability of multiple layers of different coatings including Nickel Barrier layers. Substrates available are Copper Clad Steel, ETP and OFHC Copper, Copper Alloys, Nickel and Nickel Alloys, and Steel. The company developed and introduced AccuForm® lead wires to the market years ago. AccuForm® lead wires are produced utilizing TSW's proprietary Copper Clad Steel Technology, assuring a base material of uniform and consistent metallurgical properties offering avoidance of both age and strain hardening. Outstanding solderability is of paramount importance, and all products conform to Mil Std 202, Method 208. ISO 9001 certified.

Don Edwards
Director of Sales and Marketing
tsw@torpedowire.com
1115 Instrument Drive
Rocky Mount, NC 27856
Phone: (252) 977-3900
Fax: (252) 977-4515
www.torpedowire.com

UMICORE
Greinerstraat 14
2660 Hoboken
Belgium
(32-3) 821-7111
(32-3) 821-7100 (Fax)
Products: Copper and nickel powders.

VIOX CORP.
6701 Sixth Avenue South
Seattle, WA 98108
Phone: (206) 763-2170
Products: Specialty and electronic-grade glass materials.

RAW MATERIAL

Raw Material Suppliers to the Tantalum Capacitor Industry

A. EDELHOFF GMBH & CO. KG
Am Großen Teich 33
58640 Iserlohn
Germany
(49-0-23) 714-3800
(49-0-23) 714-3807-900 (Fax)
Products: Lead Wires.

ABM RESOURCES
Level 1, 141 Broadway
Nedlands, WA 6009
Australia
(61-89) 423-9777
(61-89) 423-9733 (Fax)
Products: Tantalum Ore.

CABOT SUPERMETALS
111 Aza-Nagayachi, Higashi Naghara
Kawahigashimachi, Aizu Wakamatsu-shi
Fukushima-ken 969 3431
Japan
(81-24) 275-2868
(81-24) 275-3175 (Fax)
Products: Supplier of Tantalum Metal Powder to the Capacitor Market. Considered one of the Top Engineered Materials Companies in the World.

CABOT SUPERMETALS

Product: Tantalum Powder

Cabot Supermetals is a global supplier of high quality tantalum and niobium products for electronics, chemical and pharmaceutical processing, energy, and aerospace. Capacitor-grade tantalum powder from Cabot is used to make high reliability for cellular phones, computers, and numerous other electronic devices. These powders have enabled tantalum capacitor manufacturers to provide better performance in the same size parts and pack more performance into smaller packages. Cabot is integrated from the mine to high purity, electronics-grade tantalum. Our Aizu, Japan refining plant has earned the trust of the most demanding users in over 30 years of operations by untiring materials research and development and steadfast product quality. Our North American refining site is the only fully integrated producer of tantalum and niobium in the U.S.

Sales - Asia
Sumitomo Shiba-Daimon Bldg. 11F
2-5-5 Shiba Daimon
Minato-ku, Tokyo 105-0012
Japan
Phone: 81 3 3434 3711
Fax: 81 3 3434 6497

Sales – Americas and Europe
1095 Windward Ridge Parkway
Suite 200
Alpharetta, GA 30005
Phone: 888-390-9430
Fax: 678-297-1498
www.tantalumcentral.com

CELTIC CHEMICALS LIMITED

Product: Manganese Nitrate

Celtic Chemicals is a global supplier of high purity Manganese Nitrate solution to the tantalum capacitor industry. Celtic's high purity Manganese Nitrate is a key raw material in Tantalum capacitor production where critical applications rely on Manganese Nitrate with ultra-low impurities. "Using high purity raw materials, rigorous in-process controls, and unrivalled technical expertise, we produce Manganese Nitrate to custom specifications for our customers' critical electronic applications—consumer, medical, aerospace, and military."
For over a quarter of a century, Celtic Chemicals has provided the capacitor market's most prominent vendors with a highly reliable product that consistently excels in capacitor reliability tests. All products are RoHS and REACH compliant.

Ben Donald / Rhys Woolcock
sales@celticchemicals.co.uk
Unit 25, Kenfig Industrial Estate, Margam
Port Talbot, West Glamorgan, SA13 2PE
United Kingdom
Phone: +44 (0) 1656 749358
www.celticchemicals.co.uk

CROSSLINK ENERGY MATERIALS
950 Bolger Court
St. Louis, MO 63026
(877) 456-5864
Products: Conductive polymer cathode materials.

GIPPSLAND, LTD.
Suite 4, 207 Stirling Highway
Claremont WA 6010
Australia
(61-89) 340-6000
(61-89) 340-6060 (Fax)
Products: Tantalum Ore.

RAW MATERIAL

H.C. STARCK GMBH
Im Schleeke 78-91
38642 Goslar
Germany
(49-5) 321-7510
Products: Tantalum Cap Grade Powders, Tantalum Wire, Tantalum Furnace Materials, BAYTRON® Conductive Polymers for Capacitors.

Other plant locations:

H.C. STARCK, INC.
45 Industrial Place
Newton, MA 02161-1951
(617) 630-5800
(617) 630-5879 (Fax)
Products: Tantalum Cap Grade Powders, Tantalum Wire, Tantalum Furnace Materials, BAYTRON® Conductive Polymers for Capacitors

STARCK LTD.
1-30-5 Hamamatsucho
Minato-ku
Tokyo 105-0013
Japan
(81-35) 776-5001
Products: Tantalum Cap Grade Powders, Tantalum Wire, Tantalum Furnace Materials, BAYTRON® Conductive Polymers for Capacitors.

H.C. STARCK (THAILAND) CO., LTD.
5, I-3A Road Map Ta Phut
Industrial Estate
Rayong 21150
Thailand
(66-03) 868-3077
(66-03) 868-3043 (Fax)
Products: Tantalum Cap Grade Powders, Tantalum Wire, Tantalum Furnace Materials, BAYTRON® Conductive Polymers for Capacitors
® = A Registered Trademark of Bayer AG, Leverkusen, Federal Republic of Germany, Licensed to H.C. Starck GmbH.

JOHNSON MATTHEY SILVER & COATING TECHNOLOGIES

Product: Inks for Use in the Manufacture of Passive Components

Johnson Matthey Silver & Coating Technologies is a world class supplier of inks for use in the manufacture of passive components. Building on Johnson Matthey's expertise in precious metals, Silver & Coatings Technologies has developed a comprehensive range of RoHS compliant inks for the manufacture of high performance passive components including the latest generations of Tantalum Capacitors, MLCC, and MLV.
For Tantalum Capacitors, new high performance carbon and silver inks are available for both Manganese Dioxide and the latest Polymer Electrolyte-based capacitors. A range of high performance solderable and platable terminations is available for MLCC and MLV. Full product details can be found on our Web site at www.jmsilver.co.uk.
These high performance materials are manufactured at our facility in the United Kingdom, under the ISO 9001:2000 quality system. All products are supported globally by dedicated technical support and sale teams.

jmsilver@matthey.com
Phone: +44 (0)1763 253393
www.jmsilver.co.uk

MALAYSIA SMELTING CORP. SDN BHD
27 Jalan Pantai
P.O. Box 2
Butterworth 12700
Malaysia
(604) 333-3500
(604) 332-6499 (Fax)
Products: Extractors of Tantalum from Tin Slag Operations.

METALYSIS, UK
Unit 2, Farfield Park
Manvers Way
Wath upon Dearne, Rotherham
South Yorkshire S63 5DB
United Kingdom
(440-170) 987-2111
(440-170) 987-1222 (Fax)

NAC KAZATOMPROM
168, Bogenbai batyr St
Almaty, 050012
Republic of Kazakhstan
Ukraine
(77-27) 261-5425

NINGXIA NON-FERRIS METALS SMELTERY
P.O. Box 105
Shizuishan City
Ningxia 753000
China
(86-95) 209-8888
(86-95) 201-2018 (Fax)
Products: Tantalum Powder (Anode) and Wire.

PROTAVIC AMERICA, INC.
8 Ricker Ave.
Londonderry, NH 03053
(800) 807-2294
(978) 372-2016 (Int'l)
Products: Raw Material Supplier to Ceramic Capacitor and Tantalum Capacitor Industries.

PURE MATERIAL LABORATORIES (PML)
5-30-1 Kumegawa-cho
Higashi-Mura-Yama
Tokyo 189
Japan
(81-42) 394-4501
Products: Supplier of Tantalum Anodes in Japan.

SOLIKAMSK MAGNESIUM WORKS
St. Pravda 9
618541 Solikamsk
Russia
Products: A processor of Columbium and Tantalum Concentrates to produce Oxides, Pentachlorides, and Lithium Salts.

TANTALUM MINING CORP. OF CANADA, LTD. (TANCO)
Subsidiary of Cabot Corp.
Box 2000
Lac du Bonnet
Manitoba ROE 1A0
Canada
(204) 884-2400
(204) 884-2211 (Fax)
Products: Tantalum Mining Operations owned by Cabot Supercapacitors since 2002.

THAILAND SMELTING & REFINING CO., LTD.
80 Moo 8 Sakdidej Road
Tambon Vichit, Amphur Muang
Phuket 83000
Thailand
(66-07) 639-1111
(66-07) 639-1120 (Fax)
Products: Recovers Tantalum From Tin Mining Operations.

RAW MATERIAL

TANTALUM PELLET COMPANY

Products: Tantalum Anodes, Tantalum Cathodes, Anode and Cathode Oxide Formation

We specialize in anodes and cathodes for wet tantalum capacitors. We have years of experience making large and irregularly shaped anodes to exacting tolerances. We can also do oxide formation on all parts we make.

Todd Knowles
President
Sales@tantalum-pellet.com
21421 N. 14th Avenue
Phoenix, AZ 85027
Phone: (623) 582-5555
www.tantalum-pellet.com

TORPEDO SPECIALTY WIRE, INC
1115 Instrument Drive
Rocky Mount, NC 27804
(252) 977-3900
(252) 977-4515 (Fax)
Products: Electronic Lead Wires.

ULVAC MATERIALS, INC.
516 Yokota, Sammu-shi,
Chiba-ken 289-1297,
Japan
(81-47) 589-0151
(81-47) 589-1469 (Fax)

RAW MATERIAL

Raw Material Suppliers to the Aluminum Electrolytic Capacitor Industry

ALCAN
725, rue Aristide Berges
Voreppe 38341
France
(33-047) 657-8000
Products: Electrolytic foil for the capacitor industry and etching services for the aluminum electrolytic capacitor industry.

ALCAN SINGEN GMBH
Alusingen-Platz 1
G-78224 Singen/Htwl.
Germany
(49) 773-1800
Products: Supplies Etched Aluminum Foil for the Aluminum Electrolytic Capacitor Industry. Products are distributed by Winter-Wolff International.

ALCOA BADIN WORKS
Highway 740
P.O. Box 576
Badin, NC 28009
United States
(704) 422-3621
Products: Anodes and Cathodes.

BCCOMPONENTS (VISHAY)
Ebentaler Str. 140
9020 Klagenfurt
Austria
Products: Etched Anode and Cathode Foils (Captive).

BECROMAL SPA
Via E. Ch. Rosenthal 5
20089 Quinto Stampi
Rozzano (Milano)
Italy
(39-0) 289-2131
Products: Etched Aluminum Foils and key supplier of related Capacitor materials.

BOYD CONVERTING CO., INC.
501 Pleasant St.
P.O. Box 287
South Lee, MA 01260
(413) 243-2200
(413) 243-4460 (Fax)
Products: Sheet Converter and Slitter.

CHEMI-CON YAMAGATA CORP.
1-1, Saiwai-cho
Nagai, Yamagata 993-8511
Japan
(81-23) 884-2131
(81-23) 884-2396 (Fax)
Products: Etched Anode and Cathode Foils in the Merchant Market.
A major capacitor supplier and subsidiary of Nippon Chemi-Con.

CROSSLINK ENERGY MATERIALS
950 Bolger Court
St. Louis, MO 63026
(877) 456-5864
Products: Conductive polymer materials.

GLATFELTER
Suite 500
96 S. George Street
York, PA 17401
(866) 744-7380
Products: Supplier of Separator Paper.

H.C. STARCK GMBH
Im Schleeke 78-91
38642 Goslar
Germany
(49-5) 321-7510
Products: BAYTRON® Conductive Polymers For Capacitors.

Other locations:

H.C. STARCK INC.
45 Industrial Place
Newton, MA 02161-1951
(617) 630-5800
Products: BAYTRON® Conductive Polymers for Capacitors.

H.C. STARCK LTD.
1-30-5 Hamamatsucho
Minato-ku
Tokyo 105-0013
Japan
(81-35) 776-5001
Products: BAYTRON® Conductive Polymers for Capacitors.
® = a registered trademark of Bayer AG, Leverkusen, Federal Republic of Germany, licensed to H.C. Starck GmbH.

H.C. STARCK CO., LTD.
5, I-3A Road, Map Ta Phut
Industrial Estate
Rayong 21150
Thailand
(66-38) 683-077

HYDRO ALUMINUM IN NORTH AMERICA
801 International Drive, Suite 200
Linthicum, MD 21090
410-487-4500
410-487-8053 (Fax)
Products: Etched Anode and Cathode Foils.

JAPAN PULP & PAPER, LTD.
6-11, Nihonbashi Hongoku-cho 4 chome
Chuo-ku Tokyo 103-8641
Japan
(81-35) 201-6264
(81-33) 245-8590 (Fax)
Products: Suppliers of Separator Paper.

MH TECHNOLOGIES
2724 Park Drive
Parma, OH 44134
(216) 323-7246
(773) 496-7906 (Fax)
Products: Kraft and Electrolytic Grade Paper for the Aluminum Electrolytic Industry. Former Kimberly-Clark Plant.

NIPPON LIGHT METAL CO., LTD.
Tennozu Yusen Bldg.
2-2-20 Higashi-shinagawa
Shinagawa-ku, Tokyo 140-8628
Japan
(81-35) 461-9211
Products: Thin Aluminum for Capacitors.

NKK-NIPPON KIDOSHI
648 Hirooka-kami, Haruno-Cho
Kochi-City, Kochi, 781-0395
Japan
(81-88) 894-2321
(81-88) 894-5401 (Fax)
Products: Electrolytic Foil for Capacitors.

NORCOMP
3300 International Airport Drive, Suite #200
Charlotte, N.C. 28208
(800) 849-4450
(704) 424-5648 Fax
Products: Cans and Cases for the Aluminum Electrolytic Capacitor Industry.

OKB TITAN

Product: Black Cathode WFC Foil

OKB TITAN manufactures cathode WFC foil by using a vacuum electron beam technology of Ti evaporation for its coating deposition onto both sides of Al-foil of 20-30 micron thickness. High-porous Ti & TiN coating of 0.3 – 0.8 micron thickness provides a substantial increase of Al-foil surface area and capacity (up to 2,000 uF/cm²). Increased capacity of WFC foil in full realizes anode foil capacity in aluminum high, medium and low voltage capacitors as well as in new conductive polymer electrolyte capacitors. In contact with electrolytes, WFC foil shows a higher corrosion resistance, which results in extended life time of capacitors under critical pulse and frequency load. Cl ions content in WFC foil is null. Subject to its application, WFC foil is manufactured with capacity ranging from 400 uF/cm² to 2,000 uF/cm² and in compliance with a customer specification.

Lena Solovey
Manager
lena@okbtitan.ru
info@okbtitan.ru
109147 Russia, Moscow
Marksistskaya str., 34, building 4
Phone: +7 495 911 03 29; +7 495 912 08 16
www.okbtitan.ru

ORMET CORP.
P.O. Box 176
Hannibal, OH 43931
(740) 483-1381
(740) 483-2622 (Fax)
Products: Aluminum Sheet, Plate, and Foil.

RAW MATERIAL

PANASONIC ELECTRONIC DEVICES, CO., LTD.
1006 Oazo Kadoma
Kadoma City
Osaka 571-8506
Japan
(81-66) 906-1652
Products: MLCCs, MLC Leaded, and SLCs.
Panasonic is primarily known as a blanket supplier of both fixed and variable on-board capacitors; they supply as much product by type, configuration, and dielectric as they possibly can on a worldwide basis. Panasonic has been widely successful in DC film capacitors outside of Japan. They also produce ceramic capacitors and tantalum capacitors for captive consumption and merchant market sales.

PANASONIC ELECTRONIC DEVICES CO., LTD.
5105 National Dr.
Knoxville, TN 37914
(865) 673-0700
(865) 673-0309 (Fax)
Products: Foil production for consumption in Large Can Electrolytics and AC Power Film Capacitors.

PLASTECH GROUP LTD

Products: Screw Terminal Covers for Aluminium Electrolytic Capacitors

Plastech Group Ltd is Europe's market leader in the manufacture of high quality Injection Mouldings to the Aluminium Electrolytic Capacitor Industry.
- Standard Range of Screw Terminal Capacitor Tops (Covers/Decks) from 33 – 100 mm Diameter
- Specialist Thermoset & Thermoplastic Materials developed to withstand the demands of modern industrial Screw Terminal Capacitor applications
- Comprehensive Range of Terminal Types, Sealing Ring Gaskets & Vent Plugs
- Complete Product Design Facilities for New Products and Specials
- Fully Accredited to BS EN ISO 9001:2000 Quality Management Systems
- One Stop Solution for Performance Capacitor Components

All Capacitor Components are manufactured to exacting quality standards from the highest quality materials. Backed by our unbeatable technical support and design expertise, you won't find a better solution.

Tom Campbell
Operations Director
tom.campbell@plastechgroup.com
+44 (0) 1592 630100
Plastech Group LTD.
Moulding Division
20 Faraday Road
Southfield Industrial Estate
Glenrothes, Fife
KY6 2RU
Scotland
www.plastechgroup.com

SHINE-TOP GROUP
1 Xingtong Road, Dongjiang Precinct
Chenjiang Town, Huiyang
Guangdong
China
(86-752) 389-3888
(86-752) 389-3838 (Fax)
Products: Etched Cathode Foil, Etched and Formed Aluminum Foil for Electrolytic Capacitors.

SHOWA ALUMINUM POWDER K.K.
410 Gose, Muro
Nara Prefecture 639-2277
Japan
(81-74) 562-8266
(81-74) 565-1088 (Fax)
Products: Thin Aluminum Foil for the Capacitor industry. Aluminum Rolling and Drawing.

SPEZIALPAPIERFABRIK OBERSCHMITTEN GMBH
Rhonstr. 13
63667 Nidda / Ober-Schmitten
Germany
(49-60) 438-0801
Products: Electrolytic Grade Paper.

SUMITOMO METAL MINING CO., LTD.
11-3 Shimbashi 5-chome
Minato-ku, Tokyo 105-8716
Japan
(81-33) 436-9700
Products: Aluminum Sheet, Plate, and Foil.

SYCON POLYMERS INDIA PVT. LTD

Product: Thiophene Derivatives

Sycon provides high quality thiophene-based materials for potential use in polymeric electronics, including:

- Ethylene Dioxythiophene Liquid
- Propylene Dioxythiophene Powder
- 2,5-Dimethoxythiophene Powder
- Other Novel Thiophene Derivatives

Mr. Atul Birla
Marketing Director
info@sycon.in
R-422, MIDC, Rabale
Navi Mumbai, 400 701, INDIA
Phone: 91-22-2769 1192
www.sycon.in

WINTER-WOLFF INTERNATIONAL, INC.
131 Jericho Turnpike
Jericho, NY 11753
(516) 997-3300
(516) 997-3016 (Fax)
Products: Distributor of raw materials to the Capacitor industry for Electrolytic Capacitor Papers and Etched Aluminum Foils.

RAW MATERIAL

Raw Material Suppliers to the Film Capacitor Industry

A. EDELHOFF GMBH
Am Großen Teich 33
58640 Iserlohn
Germany
(49-0-23) 714-3800
(49-0-23) 714-3807-900 (Fax)
Products: Lead Wires.

ALCAN PACKAGING KREUZLINGEN AG
Finkernstrasse 34
CH-8280 Kruezlingen
Switzerland
(41-71) 677-7111
Products: Aluminum Foils for use in Film/Foil Capacitors.

APPLIED MATERIALS
Siemensstrasse 100
D-63755 Alzenau
Germany
(49-602) 392-6000
Products: High vacuum metallizer for film capacitor applications.

ARCOTRONICS GROUP (KEMET)
Via San Lorenzo, 19
40037 Sasso Marconi
Bologna
Italy
(39-05) 193-9111
(39-05) 184-0684 (Fax)
Products: Capacitor Winding Machines and Finished DC Film Capacitors.

BIRKELBACH KONDENSATORTECHNIK GMBH

Products: Technical Films for Capacitors

Birkelbach Kondensatortechnik offers the full range of metallized and plain films for the capacitor industry. The product range is from 3 to 20 µm OPP, 0.7 to 75 µm PET, 1.2 to 25 µm PPS, and 1.4 to 12 µm PEN. Other dielectric films like Paper, Kapton, Teflon, or PEEK are available on request.
We metallize with aluminium, zinc, alloy, and other metals like copper, silver, or gold. The slitting width is from 2.25 to 165 mm, above upon request.
We supply our high quality films worldwide from our local factory in Germany. Our task is to be a leading specialist for capacitor films using in house developed and constructed equipment on a highest level for quality and flexibility.

Holger Birkelbach
General Manager
info@birkelbachfilm.de
holger.birkelbach@birkelbachfilm.de
Im Grünewald 4
D-57339 Erndtebrück
Phone: +49-2753-5946-0
Fax: +49-2753-5946-46
www.birkelbachfilm.de

BOLLORÉ, DIVISION FILMS PLASTIQUES
Odet Ergué-Gabéric
29556 Quimper cedex 9
France
(33-29) 866-7200
(33-29) 859-6779 (Fax)
Products: Metallized Film.

BOREALIS POLYMERS N.V.

Products: High Purity, Super Clean PP Raw Material for Capacitor Film

Borealis is a world-leading supplier of Borclean™ super clean polypropylene granules for the manufacture of state-of-the-art polypropylene capacitor films. Besides the very high purity of the granules, Borealis has tailored the polymer characteristics for efficient production of bi-axially oriented polypropylene (BOPP) films with a thickness of 3 µm or less. Borealis offers grades for hazy - or rough - film applications for high voltage power factor correction (HV PFC) and for metallizable films, with technologies ranging from standard motor run capacitors to the latest developments in Hybrid Electric Vehicles (HEV), where extreme temperature resistance is required.

Franck Jacobs
Technical Service & Market Development Engineer
franck.jacobs@borealisgroup.com
Industrieweg 148
B-3583 Beringen
Belgium
Phone: +32 11 45 94 37
www.borealisgroup.com

THE DOW CHEMICAL CO.
Advanced Materials
2030 Dow Center
Midland, MI 48674
(989) 636-1000
Products: Experimental Films for the Capacitor Industry.

DUPONT ELECTRONICS
Dupont Microcircuit Materials Division
14 T.W. Alexander Dr.
RTP, NC 27709
(800) 284-3382
Products: Supplier of Polyester Film to the global DC Film Capacitor industry.

RAW MATERIAL

GRILLO-WERKE AG
Weseler Str. 1
D-47169 Duisburg
Germany
(492) 035-5570
(49-203) 555-7440
Products: Zinc Wire Leads for Film Capacitors.

MITSUBISHI POLYESTER FILM GMBH
Kasteler Straße 45
65203 Wiesbaden
Germany
(49-06) 119-6203
(49-06) 119-62-9357
Products: Metallized Film Supplier to the global Capacitor Industry.

OPPC CO., LTD.
Sinagawa Center Bldg. 9F
3-23-17 Takanawa, Minato-ku
Tokyo 108-0074
Japan
(35) 447-6733
(35) 447-6747 (Fax)
Products: Specializing in developing and manufacturing production equipment for various Electronic Components; mainly Capacitors.

SCHLENK METALLFOLIEN GMBH & CO. KG
Barnsdorfer Hauptstr. 5
D-91154 Roth
Germany
(49-9) 171-8080
(49-917) 180-8200 (Fax)
Products: Copper Foil for Film/Foil Capacitors.

SKC, LTD.
Kycbo Tower
1303-22 Seccho 4-dong
Seccho-gu, Seoul
Korea
(02) 787-1234
(02) 567-5833 (Fax)
Products: Metallized Film for the Korean Capacitor industry.

STANNIOLFABRIK EPPSTEIN GMBH & CO. KG
Burgstrasse 81-83
65817 Eppstein
Germany
(49-06) 198-5720
(49-06-19) 857-2472 (Fax)
Products: Tin Foils for Power Transmission and Distribution Capacitors.

STEINER GMBH & CO., KG
P.O. Box 20
D-57335 Erndtebrück
Germany
(492) 753-6070
(492) 753-6071-53 (Fax)
One of the largest film metallizers in the world.

TERVAKOSKI FILMS GROUP
FL-12400 Tervakoski
Finland
(35) 819-7711
Products: Ploypropylene films for capacitors.

TORAY INDUSTRIES, INC.
Nihonbashi Mitsui Tower
1-1 Nihonbashi-Muromachi 2-chome
Chuo-ku, Tokyo 103-8666
Japan
(81-33) 245-5111
(81-33) 245-5555 (Fax)
Products: Film Metallizer.

TREOFAN GROUP

Products: Treofan PHD® High Temperature Capacitor-Grade Film for Metallisation
Treofan PGD® Capacitor-Grade Film for Metallisation
Treofan GXD® Wrapping Film for Capacitor Coils

Treofan Group offers a wide range of capacitor grade films for metallised film capacitors under the brand name Treofan PHD® and Treofan PGD® Treofan PHD® has been developed from special high isotactic resin that provides particular thermo-mechanical and electrical properties to the film, thus making it suitable for both AC and DC capacitors. With its patented design, Treofan PHD® is to be used at elevated ambient temperatures, e.g. in lighting and motor run/motor start applications. Ultra thin gauges of 3 μm (Treofan PHD® 3.0) have been approved for DC-link applications such as inverter circuits in hybrid electrical vehicles (HEVs). With Treofan PGD®, Treofan markets a newly developed film type for high quality standard applications in film capacitors. A selection of wrapping films (Treofan GXD®) to protect, solidify, and insulate capacitor coils completes the Treofan® capacitor film portfolio.

Constanze Schneider
Product Manager
constanze.schneider@treofan.com
communications@treofan.com
Treofan Group
Am Prime Parc 17
65479 Raunheim
Germany
www.treofan.com

WINTER-WOLFF INTERNATIONAL, INC.
131 Jericho Turnpike
Jericho, NY 11753 USA
(516) 997-3300
(516) 997-3016 (Fax)
Products: Distributor of Films, Foils, and Papers for the Capacitor industry. Products distributed include: Birkelbach Metallized and Plain Films; Glatz Capacitor Papers; LM Neher Foils for Film/Foil Caps; LM Singen Electrolytic Foils; Brooklyn Tin Foils; Stanniolfabrik Tin Foils; Schlenk Copper Foils; Grillo Zinc Wire; Arcotronics Winding Machines, and 2A Srl Capacitor Assembly Machines.

3M CORP.
Electrical Markets Division
A130-04-N-36
6801 River Place Blvd.
Austin, TX 78726-9000
(800) 245-3573
(800) 245-0329 (Fax)
Products: Experimental Films for the Capacitor industry.

RAW MATERIAL

Raw Material Suppliers to the EDLC Supercapacitor Industry

BATSCAP
Bollore Division Films Plastiques
Odet Ergue Gaberic
F 29 556 Quimper Cedex 9
France
(33-029) 866-7200
(33-029) 859-6779 (Fax)
Products: Raw materials for Resistors.

BOLLORE TECHNOLOGIES
Odet Ergué Gabéric
29 556 Quimper Cedex 9
France
(33-029) 866-7200
(33-029) 859-6779 (Fax)
Products: Developer of a carbon Supercapacitor Substrate Material.

CALGON CARBON CORP.
400 Calgon Carbon Drive
Pittsburgh, PA 15205
(412) 787-6700
(412) 787-4523 (Fax)
Products: Carbon Cloth for the Supercapacitor Industry.

COVALENT ASSOCIATES
921 NW 11th Street
Corvallis, OR 97330
(541) 207-3844
(541) 207-3845 (Fax)
Products: Electrolyte Manufacturer.

HONEYWELL
101 Columbia Road
Morristown, NJ 07960
(973) 455-2000
Products: High purity, organic electrolytes for use in supercapacitors.

KURARAY CHEMICAL CO., LTD.
Shin-Hankyu Building 9F
1-12-39 Umeda, Kita-ku
Osaka, 530-8611
Japan
(81-66) 348-9580
(81-66) 348-9651 (Fax)
Products: Carbon Cloth for use in EDLC's.

MEADWESTVACO
Specialty Chemicals Division
5600 Virginia Avenue
North Charleston, SC 29406
(843) 740-2300
Products: Paper Chemicals, Publication Ink Resins, Water Based Polymers.

MITSUBISHI CHEMICAL CORP.
14-1 Shiba 4-chome, Minato-ku,
Tokyo 108-0014
Japan
(81-036) 414-3000
(81-036) 414-3671
Products: Supercapacitor Electrolyte research and development.

OPPC
Sinagawa Center Bldg. 9F
3-23-17 Takanawa, Minato-ku
Tokyo 108-0074
Japan
(81-35) 447-6733
(81-35) 447-6747
Products: Manufacturer of Capacitor production systems, including Double Layer Carbon Capacitor Systems

OSAKA GAS CO., LTD.
1-2, Hiranomachi 4-chome
Chuo-ku, Osaka 541
Japan
(81-06) 205-4568
Products: Gas Company and Carbon Materials.

PICA
1 Place Montgolfier
94417 Saint Maurice Cedex
France
(33-14) 511-5400
Products: Activated Carbon Fiber Fabric Material.

RUBYCON CORP., LTD.
1938-1 Nishi-Minowa
Ina City, Nagano Prefecture
Japan
Products: Electrolytic and DC Film Capacitors.

SACHEM INC.
821 Woodward St.
Austin, Texas 78704
(512) 421-4900
(512) 445-5066 (Fax)
Products: Electrolytes.

SPECTRACORP (ENGINEERED FIBERS TECHNOLOGY)
25 Brook Street, Ste. B
Shelton, CT 06484-3177
(203) 922-1810
Products: 2500m^2/g Carbon Paper for Supercapacitor Manufacturing.

W.L. GORE & ASSOCIATES
555 Papermill Road
Newark, DE 19711
(410) 506-7787
(888) 914-4673 (U.S.)
Products: Carbon PTFE Electrode Materials and Separators.

Raw Material Suppliers of Niobium

CBMM
Companhia Brasileira de Metalurgia e Mineração
Rua Pequetita, 111-Vila Olímpia
04552-060 - São Paulo, SP
Brazil
(55-113) 371-9222
(55-113) 820-2090
Products: Largest supplier of Niobium products to the global market.

H.C. STARCK GMBH
Im Schleeke 78-91
38642 Goslar
Germany
(49-5) 321-7510
Products: Niobium, Niobium Oxide, Niobium Oxalate, Niobiates.

Other plant locations:

H.C. STARCK, INC.
45 Industrial Place
Newton, MA 02161-1951
(617) 630-5800
Products: Niobium, Niobium Oxide, Niobium Oxalate, Niobiates.

H.C. STARCK, LTD.
1-30-5 Hamamatsucho
Minato-ku
Tokyo 105-0013
Japan
(81-35) 776-5001
Products: Niobium, Niobium Oxide, Niobium Oxalate, Niobiates.

RAW MATERIAL

Raw Material Suppliers to the Resistor Industry

A. EDELHOFF GMBH
Am Großen Teich 33
58640 Iserlohn
Germany
(49-0-23) 714-3800
(49-0-23) 714-3807-900 (Fax)
Products: Lead Wires.

BASF CATALYSTS LLC
1 West Central Avenue
East Newark, NJ 07029
(973) 268-7800
(973) 268-7913 (Fax)
Products: Thick-Film Resistive Pastes.

CERAMTEC AG
Fabrikstr. 23–29
D-73207 Plochingen
Germany
(49-01) 153-6110
Products: Supplier of Alumina Cores and Substrates for the global Resistor industry.

COORSTEK INC.
16000 Table Mountain Pkwy
Golden, CO 80403
(303) 271-7000
(303) 271-7009 (Fax)
Products: One of the main suppliers of 99% Alumina Substrates for the Thick-Film Resistor Chip, Network and Array markets.

DUPONT ELECTRONICS, INC.
Dupont Microcircuit Materials Division
14 T.W. Alexander Dr.
RTP, NC 27709
(800) 284-3382
Products: Resistive Inks.

EXOJET TECHNOLOGY CORP.
6F, No. 22, Yai Yuen Street
Chupei, Hsinchu 302
Taiwan
(88-63) 552-6085
(88-63) 552-6086 (Fax)
Products: Conductive and insulating materials.

FERRO CORP.
3900 S. Clinton Avenue
S. Plainfield, NJ 07080
(908) 561-1100
Products: Thick-Film Resistive Powders and Pastes.

HERAEUS, INC.
Heraeus Circuit Materials Division
24 Union Hill Road
W. Conshohocken, PA 19428
(610) 825-6050
Products: Resistive Pastes.

HOKURIKU ELECTRIC INDUSTRY CO., LTD.
3158 Shimo-okubo
Toyama City, Toyama Pref. 939-2292
Japan
(076) 467-1111
Products: Resistor and Resistive Pastes.

JOHNSON MATTHEY
40-42 Hatton Garden
London EC1N 8EE
(0-20-7) 269-8400
(44-02-07) 269-8433 (Fax)
Products: Supplier of Refined Ruthenium Bullion.

KYOCERA INDUSTRIAL CERAMICS CORP.
5713 East Fourth Plain Blvd.
Vancouver, WA 98661
(81-75) 604-3500
Products: Supplier of Alumina Cores and Substrates.

SHOEI CHEMICAL, INC.
Shinjuku Mitsui Building
1-1, Nishi-Shinjuku 2-chome
Shinjuku-ku, Tokyo 163-0443
Japan
(81-03) 344-6661
Products: Electrode Powders and Pastes. Intermediate supplier of palladium and palladium-silver paste.

SUMITOMO METAL MINING CO., LTD.
11-3 Shimbashi 5-chome
Minato-ku, Tokyo 105-8716
Japan
(81-33) 436-9700
Products: Electrode and Termination Powders and Pastes, including both Precious and Base Metals. Also sells Resistive Powders and Pastes.

TANAKA KIKINZOKU INTERNATIONAL K.K.
Marunouchi Trust Tower N-12F
1-8-1 Marunouchi, Chiyoda-ku, Tokyo 100-0005
Japan
(81-35) 222-1380
Products: Supplier of Ruthenium Oxide Resistive Inks and Palladium Silver Resistive Inks.

TORPEDO SPECIALTY WIRE, INC.

Product: Electronic Lead Wires

Torpedo Specialty Wire is a leading manufacturer of electronic lead wires for the capacitor and resistor industry for more than 34 years. The company has led the way in developing new lead wire products and processes to meet the customer's needs. Product line consists of a variety of coatings including Tin/Lead Solders of all compositions, 100% Tin-RoHS compliant, and Nickel with the capability of multiple layers of different coatings including Nickel Barrier layers. Substrates available are Copper Clad Steel, ETP and OFHC Copper, Copper Alloys, Nickel and Nickel Alloys, and Steel. The company developed and introduced AccuForm® lead wires to the market years ago. AccuForm® lead wires are produced utilizing TSW's proprietary Copper Clad Steel Technology, assuring a base material of uniform and consistent metallurgical properties offering avoidance of both age and strain hardening. Outstanding solderability is of paramount importance, and all products conform to Mil Std 202, Method 208. ISO 9001 certified.

Don Edwards
Director of Sales and Marketing
tsw@torpedowire.com
1115 Instrument Drive
Rocky Mount, NC 27804
USA
Phone: (252) 977-3900
Fax: (252) 977-4515
www.torpedowire.com

The Electronic Components Association (ECA) is a non-profit trade association dedicated to supporting the needs and interests of the global supply chain for the manufacturing and suppliers of passive and electro-mechanical electronic components, connectors, wire and cable, component arrays and assemblies, and materials and support services.

ECA
Electronic Components Association

ECA MEMBERS HAVE

The best face-to-face marketing, meeting and technical information opportunities

Access to millions of potential customers over the web

The ability to influence world-wide technology challenges

A voice on the issues that shape and define the market

www.ecaus.org

www.ingramcontent.com/pod-product-compliance
Lightning Source LLC
Chambersburg PA
CBHW051050180526
45172CB00002B/589